D1219135

FINITE REFLECTION GROUPS

FINITE REFLECTION GROUPS

C. T. BENSON
University of Oregon

L. C. GROVE
Syracuse University

BOGDEN & QUIGLEY, INC.
PUBLISHERS

Tarrytown-on-Hudson, New York / Belmont, California

Cover designed by Winston G. Potter

Text designed by Science Bookcrafters, Inc.

L. C. Catalog Card No.: 74-150337

Standard Book No.: -0-8005-0001-6

Printed in the United States of America

1 2 3 4 5 6 7 8 9 10—75 74 73 72 71

PREFACE

This book began as lecture notes for a course given at the University of Oregon. The course, given for undergraduates and beginning graduate students, follows immediately after a conventional course in linear algebra and serves two chief pedagogical purposes. First, it reinforces the students' newly won knowledge of linear algebra by giving applications of several of the theorems they have learned and by giving geometrical interpretations for some of the notions of linear algebra. Second, some students take the course before or concurrently with abstract algebra, and they are armed in advance with a collection of fairly concrete nontrivial examples of groups.

The first comprehensive treatment of finite reflection groups was given by H. S. M. Coxeter in 1934. In [6] he completely classified the groups and derived several of their properties, using mainly geometrical methods. He later included a discussion of the groups in his book *Regular Polytopes* [7]. Another discussion, somewhat more algebraic in nature, was given by E. Witt in 1941 [27]. An algebraic account of reflection groups was presented by P. Cartier in the Chevalley Seminar reports (see [5]). Another has recently appeared in N. Bourbaki's chapters on Lie groups and Lie algebras [3].

Since the sources cited above do not seem to be easily accessible to most undergraduates, we have attempted to give a discussion of finite reflection groups that is as elementary as possible. We have tried to reach a middle ground between Coxeter and Bourbaki. Our approach is algebraic, but we have retained some of the geometrical flavor of Coxeter's approach.

Chapter 1 introduces some of the terminology and notation used later and indicates prerequisites. Chapter 2 gives a reasonably thorough account of *all* finite subgroups of the orthogonal groups in two and three dimensions. The presentation is somewhat less formal than in succeeding chapters. For instance, the existence of the icosahedron is accepted as an empirical fact, and no formal proof of existence is included. Throughout most of Chapter 2 we do not distinguish between groups that are "geometrically indistinguishable," that is, conjugate in the orthogonal group. Very little of the material in Chapter 2 is actually required for the subsequent chapters, but it serves two important purposes: It aids in the development of geometrical insight, and it serves as a source of illustrative examples.

There is a discussion of fundamental regions in Chapter 3. Chapter 4 provides a correspondence between fundamental reflections and fundamental regions via a discussion of root systems. The actual classification and construction of finite reflection groups takes place in Chapter 5, where we have in part followed the methods of E. Witt and B. L. van der Waerden. Generators and relations for finite reflection groups are discussed in Chapter 6. There are historical remarks and suggestions for further reading in a Postlude.

Since we have written with the student in mind we have included considerable detail and a number of illustrative examples. Exercises are included in every chapter but the first. The results of some of the exercises are used in the body of the text. The list of identifications in Exercise 5.7 was worked out by one of our students, Leslie Wilson.

We wish to thank James Humphreys, Otto Kegel, and Louis Solomon for reading the manuscript and making numerous excellent suggestions. We also derived considerable benefit from Charles Curtis's lectures on root systems and Chevalley groups.

C.T.B. AND L.C.G.

July 1970

CONTENTS

FINITE REFLECTION GROUPS

chapter 1

PRELIMINARIES

1.1 LINEAR ALGEBRA

We assume that the reader is familiar with the contents of a standard course in linear algebra, including finite-dimensional vector spaces, subspaces, linear transformations and matrices, determinants, eigenvalues, bilinear and quadratic forms, positive definiteness, real inner product spaces, and orthogonal linear transformations. Accounts of these topics may be found in most linear algebra books (e.g., [11] or [17]). Throughout the book V will denote a real Euclidean vector space, i.e., a finite-dimensional inner product space over the real field $\mathscr{R}$. Partly in order to establish notation we list some of the properties of V that are of importance for the ensuing discussion.

If X and Y are subsets of V such that $(x, y) = 0$ for all $x \in X$ and all $y \in Y$, we shall say that X and Y are *orthogonal*, or *perpendicular*, and write $X \perp Y$. If $X \subseteq V$, the *orthogonal complement* of X, which is the subspace of V consisting of all $x \in V$ such that $x \perp X$, will be denoted by $X^{\perp}$. If W is a subspace of V, then $W^{\perp\perp} = W$ and $V = W \oplus W^{\perp}$.

If $\{x_1, \ldots, x_n\}$ is a basis for V, let V_i be the subspace spanned by $\{x_1, \ldots, x_{i-1}, x_{i+1}, \ldots, x_n\}$, excluding x_i. If $0 \neq y_i \in V_i^{\perp}$, then $(x_j, y_i) = 0$ for all $j \neq i$, but $(x_i, y_i) \neq 0$, for otherwise $y_i \in V^{\perp} = 0$. Dividing y_i by (x_i, y_i), if necessary, we may assume that $(x_i, y_i) = 1$, thereby making y_i unique since $\dim(V_i^{\perp}) = 1$. Observe that if $\Sigma_{i=1}^{n} \lambda_i y_i = 0$ with $\lambda_i \in \mathscr{R}$, then

$$0 = (x_j, 0) = (x_j, \Sigma_i \lambda_i y_i) = \Sigma_i \lambda_i (x_j, y_i) = \lambda_j$$

for all j, and so $\{y_1, \ldots, y_n\}$ is linearly independent. Thus $\{y_1, \ldots, y_n\}$ is a basis, called the *dual basis* of $\{x_1, \ldots, x_n\}$. It is the unique basis with the property that

$$(x_i, y_j) = \delta_{ij} = \begin{cases} 1 & \text{if } i = j, \\ 0 & \text{if } i \neq j. \end{cases}$$

The space of all n-tuples (column vectors) of real numbers will be denoted by $\mathscr{R}^n$. Since there is seldom any chance of confusion we shall often write the elements of $\mathscr{R}^n$ as row vectors $(\lambda_1, \ldots, \lambda_n)$ for the sake of typographical convenience. The usual basis vectors along the positive coordinate axes in $\mathscr{R}^n$ will be denoted by

$$e_1 = (1, 0, \ldots, 0), \qquad e_2 = (0, 1, 0, \ldots, 0),$$

etc. The space $\mathscr{R}^n$ is an inner product space, with

$$((\lambda_1, \ldots, \lambda_n), (\mu_1, \ldots, \mu_n)) = \Sigma_{i=1}^n \lambda_i \mu_i.$$

If V is any real Euclidean vector space, then it is a consequence of the Gram–Schmidt theorem ([11], p. 108) that there is an inner product preserving isomorphism from V onto $\mathscr{R}^n$, where $n = \dim V$. Thus when it is convenient we shall lose no generality if we assume that $V = \mathscr{R}^n$.

The *length* $\sqrt{(x, x)}$ of a vector $x \in V$ will be denoted by $\|x\|$. If $x, y \in V$, then the *distance* between them, denoted by $d(x, y)$, is defined to be $\|x - y\|$. For a fixed vector $x_0 \in V$ and real number $\varepsilon > 0$ the set

$$\{x \in V : d(x, x_0) = \varepsilon\}$$

is called the *sphere* of radius ε centered at x_0, and the set

$$\{x \in V : d(x, x_0) < \varepsilon\}$$

is called the (open) *ball* of radius ε centered at x_0.

A subset U of V is called *open* if and only if given any $x \in U$ there is some $\varepsilon > 0$ for which the ball of radius ε centered at x lies entirely within U. The conditions of the definition are vacuously satisfied by the empty set $\varnothing$, so $\varnothing$ is open by default. Note that finite intersections and arbitrary unions of open sets are open. A subset D of V is called *closed* if and only if its complement $V \setminus D$ is open, so finite unions and arbitrary intersections of closed sets are closed. The intersection of all closed sets containing a set X is called the *closure* of X and is denoted by X^-. The *interior* X^0 of X is the union of all open subsets of X. The *boundary* of X is defined to be $X^- \setminus X^0$. For example, the sphere of radius ε centered at x_0 is the boundary of the ball with the same radius and center.

If X is a fixed subset of V and $Y \subseteq X$, then Y is called *relatively open* in X if and only if $Y = X \cap U$ for some open subset U of V. Likewise,

Y is *relatively closed* in X if and only if $Y = X \cap D$ for some closed subset D of V, and the (*relative*) *closure* of Y in X is the intersection of X with the closure Y^- of Y in V. A subset X of V is *connected* if and only if it is not the disjoint union of two nonempty relatively open subsets. At the opposite extreme X is *discrete* if and only if every point of X is a relatively open set.

If $\dim V = n$, then a *hyperplane* in V is an $(n - 1)$-dimensional subspace. A *line* in V is any translate of a one-dimensional subspace. Thus a line is a subset of the form $\{x + \lambda y : \lambda \in \mathscr{R}\}$, where x and y are fixed vectors with $y \neq 0$. The *line segment* $[xy]$ between two vectors x and y of V is the set

$$\{x + \lambda(y - x) : 0 \leq \lambda \leq 1\}.$$

Note that if $x \neq y$, then $[xy]$ is the smallest connected subset of the line

$$\{x + \lambda(y - x) : \lambda \in \mathscr{R}\}$$

that contains x and y. A subset X of V is called *convex* if and only if the line segment $[xy]$ lies wholly within X for all points x and y of X. Observe that a convex set is connected.

A *transformation* of V is understood to be a linear transformation. The group of all orthogonal transformations of V will be denoted by $\mathcal{O}(V)$. If $T \in \mathcal{O}(V)$ then $\det T = \pm 1$, and if a (complex) number λ is an eigenvalue of T then $|\lambda| = 1$. If $T \in \mathcal{O}(V)$ and $\det T = 1$, then T will be called a *rotation*.

1.2 GROUP THEORY

We shall assume that the reader is familiar with the following notions from elementary group theory: subgroup, coset, order, index, homomorphism, kernel, normal subgroup, isomorphism, and direct product. A discussion may be found in any book on group theory or almost any book on abstract algebra (e.g., [16], [1], or [19]).

If $\mathscr{S}$ is a set, the cardinality of $\mathscr{S}$ will be denoted by $|\mathscr{S}|$. In particular, the order of a group $\mathscr{G}$ is $|\mathscr{G}|$. If $\mathscr{S}$ is a subset of a group $\mathscr{G}$, then $\langle \mathscr{S} \rangle$ will denote the subgroup of $\mathscr{G}$ generated by $\mathscr{S}$. If $\mathscr{H}$ is a subgroup of $\mathscr{G}$ we write $\mathscr{H} \leq \mathscr{G}$, and $[\mathscr{G} : \mathscr{H}]$ will denote the index of $\mathscr{H}$ in $\mathscr{G}$.

A *permutation* of a set $\mathscr{S}$ is a 1-1 function from $\mathscr{S}$ onto $\mathscr{S}$. The set $\mathscr{P}(\mathscr{S})$ of all permutations of $\mathscr{S}$ is a group under the operation of composition of functions; i.e., $(fg)(x) = f(g(x))$, all $x \in \mathscr{S}$. If $\mathscr{S} = \{1, 2, \ldots, n\}$, then the group $\mathscr{P}(\mathscr{S})$ is called the *symmetric group* on n letters and is denoted by $\mathscr{S}_n$. We shall assume known the elementary properties of $\mathscr{S}_n$

(see [19], pp. 64–68). In particular, $\mathcal{S}_n$ has a subgroup of index 2, the *alternating group* on n letters, consisting of all the even permutations in $\mathcal{S}_n$.

If $\mathcal{S}$ is a set, then a group $\mathcal{G}$ is said to be (represented as) a *permutation group* on $\mathcal{S}$ if and only if there is a homomorphism φ from $\mathcal{G}$ to $\mathcal{P}(\mathcal{S})$. If φ is an isomorphism into $\mathcal{P}(\mathcal{S})$, then $\mathcal{G}$ is said to be represented *faithfully* or to be a *faithful* permutation group on $\mathcal{S}$. Note that if $\mathcal{G}$ is faithful and $\mathcal{S}$ is finite, then $\mathcal{G}$ is isomorphic with a subgroup of $\mathcal{S}_n$, and in particular $\mathcal{G}$ is finite.

If $\mathcal{G}$ is a permutation group on $\mathcal{S}$, we shall write simply Tx rather than $(\varphi T)x$ for all $T \in \mathcal{G}$, $x \in \mathcal{S}$. If $x \in \mathcal{S}$, then the subset $\mathcal{H}$ of $\mathcal{G}$, consisting of all $T \in \mathcal{G}$ for which $Tx = x$, is a subgroup called the *stabilizer* of x, denoted by $\operatorname{Stab}(x)$. The subset of $\mathcal{S}$ consisting of all Tx, as T ranges over $\mathcal{G}$, is called the *orbit* of x, denoted by $\operatorname{Orb}(x)$. If $\operatorname{Orb}(x) = \mathcal{S}$ for each $x \in \mathcal{S}$, then $\mathcal{G}$ is said to be *transitive* on $\mathcal{S}$.

Proposition 1.2.1

If $\mathcal{G}$ is a permutation group on a set $\mathcal{S}$ and $x \in \mathcal{S}$, then $[\mathcal{G} : \operatorname{Stab}(x)] = |\operatorname{Orb}(x)|$.

Proof

Set $\mathcal{H} = \operatorname{Stab}(x)$. If $R, T \in \mathcal{G}$ and $R\mathcal{H} = T\mathcal{H}$, then $T^{-1}R \in \mathcal{H}$, or $T^{-1}Rx = x$; so $Rx = Tx$. Thus $\theta(T\mathcal{H}) = Tx$ defines a mapping θ from the set of left cosets of $\mathcal{H}$ onto the orbit of x. If $Rx = Tx$, then $T^{-1}R \in \mathcal{H}$, and $R\mathcal{H} = T\mathcal{H}$. Thus θ is also 1-1 and the proposition is proved.

chapter 2

FINITE GROUPS IN TWO AND THREE DIMENSIONS

2.1 *ORTHOGONAL TRANSFORMATIONS IN TWO DIMENSIONS*

If $T \in \mathcal{O}(\mathcal{R}^2)$, then T is completely determined by its action on the basis vectors $e_1 = (1, 0)$ and $e_2 = (0, 1)$. If $Te_1 = (\mu, \nu)$, then $\mu^2 + \nu^2 = 1$ and $Te_2 = \pm(-\nu, \mu)$, since T preserves length and orthogonality. Choose $\theta, 0 \leq \theta < 2\pi$, such that $\cos\theta = \mu$ and $\sin\theta = \nu$.

If $Te_2 = (-\nu, \mu)$, then T is represented by the matrix

$$A = \begin{bmatrix} \mu & -\nu \\ \nu & \mu \end{bmatrix} = \begin{bmatrix} \cos\theta & -\sin\theta \\ \sin\theta & \cos\theta \end{bmatrix},$$

and it is clear that T is a counterclockwise rotation of the plane about the origin through the angle θ (see Figure 2.1). Observe that

$$\det T = \mu^2 + \nu^2 = \cos^2\theta + \sin^2\theta = 1.$$

If $Te_2 = (\nu, -\mu)$, then T is represented by the matrix

$$B = \begin{bmatrix} \mu & \nu \\ \nu & -\mu \end{bmatrix} = \begin{bmatrix} \cos\theta & \sin\theta \\ \sin\theta & -\cos\theta \end{bmatrix}.$$

In this case observe that

$$\det T = -\cos^2\theta - \sin^2\theta = -1,$$

and that

$$B^2 = \begin{bmatrix} \mu^2 + \nu^2 & 0 \\ 0 & \mu^2 + \nu^2 \end{bmatrix} = \begin{bmatrix} 1 & 0 \\ 0 & 1 \end{bmatrix},$$

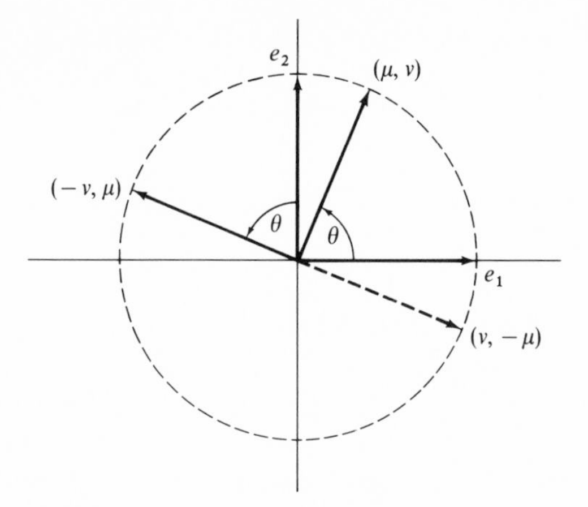

Figure 2.1

so that $T^2 = 1$. It is easy to verify (Exercise 2.1) that the vector $x_1 = (\cos\theta/2, \sin\theta/2)$ is an eigenvector having eigenvalue 1 for T, so that the line $l = \{\lambda x_1 : \lambda \in \mathscr{R}\}$ is left pointwise fixed by T. Similarly, the vector $x_2 = (-\sin\theta/2, \cos\theta/2)$ is an eigenvector with eigenvalue -1, and $x_2 \perp x_1$ [see Figure 2.2(a)]. With respect to the basis $\{x_1, x_2\}$ the transformation T is represented by the matrix

$$C = \begin{bmatrix} 1 & 0 \\ 0 & -1 \end{bmatrix}.$$

If $x = \lambda_1 x_1 + \lambda_2 x_2$, then $Tx = \lambda_1 x_1 - \lambda_2 x_2$, and T sends x to its mirror

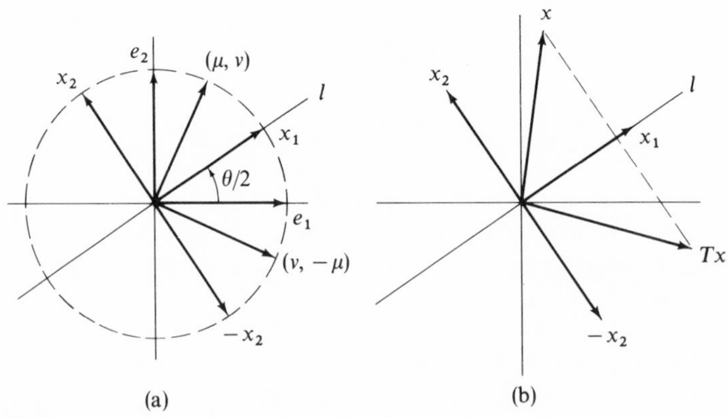

Figure 2.2

image with respect to the line l [see Figure 2.2(b)]. The transformation T is called the *reflection through* l or the *reflection along* x_2. Observe that

$$Tx = x - 2(x, x_2)x_2$$

for all $x \in \mathscr{R}^2$.

We have shown that every orthogonal transformation of $\mathscr{R}^2$ is either a rotation or a reflection.

2.2 *FINITE GROUPS IN TWO DIMENSIONS*

Suppose that $\dim V = 2$ and that $\mathscr{G}$ is a finite subgroup of $\mathscr{O}(V)$. The set of all rotations in $\mathscr{G}$ constitutes a subgroup $\mathscr{H}$ of $\mathscr{G}$. As was shown in Section 2.1, each $T \in \mathscr{H}$ is a counterclockwise rotation of V through an angle $\theta = \theta(T)$ with $0 \leq \theta < 2\pi$. If $\mathscr{H} \neq 1$, choose $R \in \mathscr{H}$ with $R \neq 1$, for which $\theta(R)$ is minimal. If $T \in \mathscr{H}$, choose an integer m such that

$$m\theta(R) \leq \theta(T) < (m + 1)\theta(R).$$

Then $0 \leq \theta(T) - m\theta(R) < \theta(R)$. But

$$\theta(T) - m\theta(R) = \theta(R^{-m}T),$$

since $R^{-m}T$ is a counterclockwise rotation through angle $\theta(T)$ followed by m *clockwise* rotations, each through angle $\theta(R)$. Since $\theta(R)$ was chosen to be minimal, we must have $\theta(R^{-m}T) = 0$; so $R^{-m}T = 1$ or $T = R^m$. In other words, $\mathscr{H} = \langle R \rangle$ is a cyclic group. It also follows that $\theta(R) = 2\pi/n$, where $n = |\mathscr{H}|$.

If $\mathscr{G} = \mathscr{H}$, we have shown that $\mathscr{G}$ is a cyclic group of order n, in which case $\mathscr{G}$ will be denoted by $\mathscr{C}_2^n$ (the subscript calls attention to the fact that $\dim V = 2$).

Suppose next that $\mathscr{G} \neq \mathscr{H}$, and choose a reflection $S \in \mathscr{G}$. Since $\det(SR^k) = \det S = -1$ for all integers k, the coset $S\mathscr{H}$ contains $n = |\mathscr{H}|$ distinct reflections. If $T \in \mathscr{G}$ is a reflection, then

$$\det(ST) = (\det S)(\det T) = (-1)(-1) = 1,$$

so $ST \in \mathscr{H}$; hence $T \in S\mathscr{H}$, since $S^{-1} = S$. Thus $\mathscr{H}$ is a subgroup of index 2 in $\mathscr{G}$, and if $\mathscr{H} = \langle R \rangle$, as above, then

$$\mathscr{G} = \langle R, S \rangle = \{1, R, \ldots, R^{n-1}, S, SR, \ldots, SR^{n-1}\},$$

and $|\mathscr{G}| = 2n$. Since RS is a reflection, we have $(RS)^2 = 1$, or $RS = SR^{-1} = SR^{n-1}$, completely determining the multiplication in $\mathscr{G}$. The group $\mathscr{G}$ is called the *dihedral group* of order $2n$, and it will be denoted by $\mathscr{H}_2^n$. We have proved

Theorem 2.2.1

If $\dim V = 2$ and $\mathscr{G}$ is a finite subgroup of $\mathscr{O}(V)$, then $\mathscr{G}$ is either a cyclic group $\mathscr{C}_2^n$ or a dihedral group $\mathscr{H}_2^n$, $n = 1, 2, 3, \ldots$.

If we set $T = RS$ in the dihedral group $\mathscr{H}_2^n = \langle S, R\rangle$, then T is a reflection, since $\det T = -1$. Since $TS = RS^2 = R$, it is clear that $\langle S, T\rangle = \mathscr{H}_2^n$, so $\mathscr{H}_2^n$ is generated by reflections. If we suppose that the orthonormal basis $\{x_1, x_2\}$ of eigenvectors of S discussed in Section 2.1 coincides with the usual basis $\{e_1, e_2\}$ in $\mathscr{R}^2$, then we may assume that S and R are represented by the matrices

$$A = \begin{bmatrix} 1 & 0 \\ 0 & -1 \end{bmatrix} \quad \text{and} \quad B = \begin{bmatrix} \cos 2\pi/n & -\sin 2\pi/n \\ \sin 2\pi/n & \cos 2\pi/n \end{bmatrix},$$

respectively. Thus T is represented by the matrix

$$C = BA = \begin{bmatrix} \cos 2\pi/n & \sin 2\pi/n \\ \sin 2\pi/n & -\cos 2\pi/n \end{bmatrix},$$

so T is a reflection through a line l inclined at an angle of π/n to the positive x-axis. Let us use these ideas to give a geometrical interpretation of the group $\mathscr{H}_2^n$.

Denote by F the open wedge-shaped region in the first quadrant bounded by the x-axis and the line l. The x-axis is a reflecting line for the transformation S, and l is a reflecting line for the transformation T. The $2n$ congruent regions in the plane obtained by rotating the region F through successive multiples of π/n can be labeled with the elements of $\mathscr{H}_2^n$ as follows: For each $U \in \mathscr{H}_2^n$, designate by U the region $U(F)$ obtained by applying U to all points of the region F.

The procedure is illustrated in Figure 2.3 for the case $n = 4$. If two plane mirrors are set facing one another along the reflecting lines for S

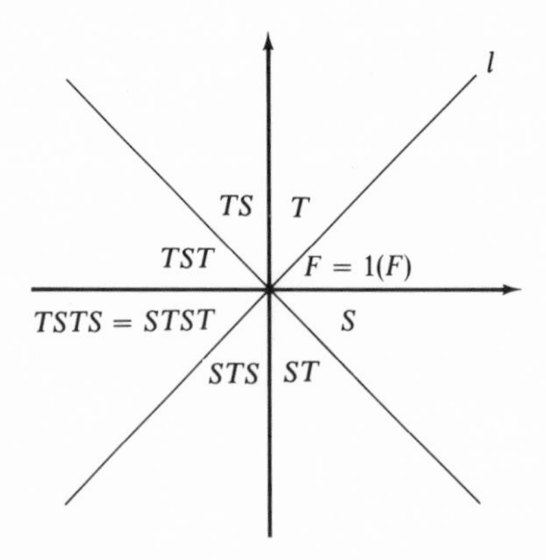

Figure 2.3

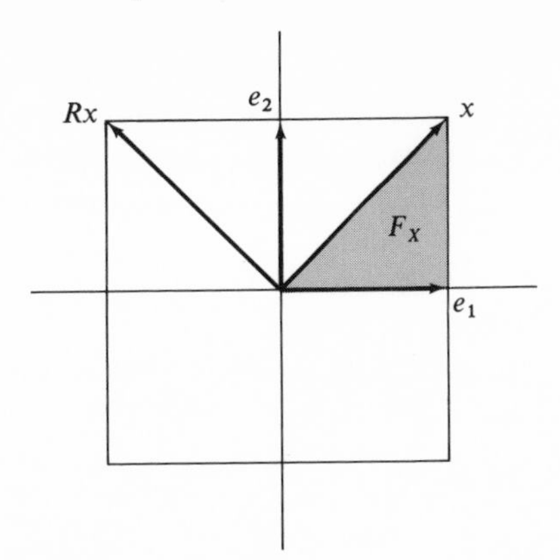

Figure 2.4

and T, with their common edge perpendicular to the plane at the origin, then the other lines may be seen in the mirrors as edges of virtual mirrors. If an object is placed between the mirrors in the region F, then reflections of the object can be seen in the seven images of F. This illustrates the principle of the kaleidoscope and shows a connection between the kaleidoscope and the dihedral groups.

Observe that the region F is open, that no point of F is mapped to any other point of F by any nonidentity element U of $\mathscr{H}_2^n$, and that the union of the closures $(UF)^-$, $U \in \mathscr{H}_2^n$, is all of $\mathscr{R}^2$. A region F with these properties will be called a *fundamental region* for the group $\mathscr{H}_2^n$. Fundamental regions will be discussed more fully in Chapter 3.

If some nonzero vector x and its image Rx under the action of the rotation R through minimal angle $\theta(R)$ are joined by a line segment, then that line segment together with its images under all transformations in $\mathscr{H}_2^n$ bound a regular n-gon X. The subgroup $\mathscr{C}_2^n$ of rotations in $\mathscr{H}_2^n$ is the group of all rotations that leave the n-gon invariant, and $\mathscr{H}_2^n$ itself is the group of all orthogonal transformations that leave X invariant. In the case $n = 4$, $\mathscr{C}_2^4$ and $\mathscr{H}_2^4$ are the rotation group and the full orthogonal group under which the square is invariant [see Figure 2.4, where $x = (1, 1)$]. The relatively open region $F_X = F \cap X$ is a fundamental region in the square X for the group $\mathscr{H}_2^4$, in the sense discussed above.

2.3 *ORTHOGONAL TRANSFORMATIONS IN THREE DIMENSIONS*

We assume throughout this section that $\dim V = 3$.

Theorem 2.3.1 (Euler)

Suppose that T is a rotation in $\mathcal{O}(V)$. Then T is a rotation about a fixed axis, in the sense that T has an eigenvector x having eigenvalue 1 such that the restriction of T to the plane $\mathcal{P} = x^\perp$ is a two-dimensional rotation of $\mathcal{P}$.

Proof

Suppose that λ_1, λ_2, and λ_3 are the eigenvalues of T. At least one of them, say λ_1, must be real, since they are the roots of a cubic polynomial with real coefficients. If λ_2 is not real, its complex conjugate is also an eigenvalue; so $\lambda_3 = \overline{\lambda_2}$. Since $\det T = \lambda_1\lambda_2\lambda_3 = 1$, the only possibilities are (relabeling if necessary)

(a) $\lambda_1 = 1, \lambda_2 = \lambda_3 = \pm 1$,

and

(b) $\lambda_1 = 1, \lambda_2 = \overline{\lambda_3} \notin \mathcal{R}$.

In either case 1 is an eigenvalue. Choose a corresponding eigenvector x and note that $x = T^{-1}Tx = T^{-1}x$. If $y \perp x$, then

$$(Ty, x) = (y, T^{-1}x) = (y, x) = 0,$$

so $\mathcal{P} = x^\perp$ is invariant under T. Since the determinant of the restriction $T|\mathcal{P}$ is the product of the other two eigenvalues λ_2 and λ_3 of T, we have $\det(T|\mathcal{P}) = 1$; so $T|\mathcal{P}$ is a rotation of the plane $\mathcal{P}$.

A *reflection* in $\mathcal{O}(V)$ is a transformation S whose effect is to map every point of V to its mirror image with respect to a plane $\mathcal{P}$ containing the origin. More precisely, S is a reflection if $Sx = x$ for all x in the plane $\mathcal{P}$, and if $Sy = -y$ for all $y \in \mathcal{P}^\perp$. If r is chosen to be a unit vector in $\mathcal{P}^\perp$, then S is given by the formula

$$Sx = x - 2(x, r)r, \qquad \text{all } x \in V.$$

If we set $x_1 = r$ and choose a basis $\{x_2, x_3\}$ for $\mathcal{P}$, then with respect to the basis $\{x_1, x_2, x_3\}$ the transformation S is represented by the matrix

$$A = \begin{bmatrix} -1 & 0 & 0 \\ 0 & 1 & 0 \\ 0 & 0 & 1 \end{bmatrix}.$$

Note that $S^2 = 1$.

Theorem 2.3.2

Suppose that $T \in \mathcal{O}(V)$ with $\det T = -1$. Then geometrically the effect of T is that of a reflection through a plane $\mathcal{P}$, followed by a rotation about the line through the origin orthogonal to $\mathcal{P}$.

Proof

If λ_1, λ_2, and λ_3 are the eigenvalues of T, then remarks similar to those in the proof of Theorem 2.3.1 show that the only possibilities are

(a) $\lambda_1 = -1, \lambda_2 = \lambda_3 = \pm 1,$

and

(b) $\lambda_1 = -1, \lambda_2 = \overline{\lambda_3} \notin \mathcal{R}.$

Choose an eigenvector x_1 corresponding to $\lambda_1 = -1$, and set $\mathcal{P} = x_1^{\perp}$. Since

$$\det(T|\mathcal{P}) = \lambda_2\lambda_3 = 1,$$

$T|\mathcal{P}$ is a rotation of the plane $\mathcal{P}$. Thus we may choose a basis $\{x_2, x_3\}$ for $\mathcal{P}$ so that the matrix representing T relative to the basis $\{x_1, x_2, x_3\}$ is

$$A = \begin{bmatrix} -1 & 0 & 0 \\ 0 & \cos\theta & -\sin\theta \\ 0 & \sin\theta & \cos\theta \end{bmatrix} = \begin{bmatrix} 1 & 0 & 0 \\ 0 & \cos\theta & -\sin\theta \\ 0 & \sin\theta & \cos\theta \end{bmatrix} \begin{bmatrix} -1 & 0 & 0 \\ 0 & 1 & 0 \\ 0 & 0 & 1 \end{bmatrix}.$$

The theorem follows.

2.4 *FINITE ROTATION GROUPS IN THREE DIMENSIONS*

Suppose that $\dim V = 3$ and that W is a plane in V, i.e., a subspace of dimension 2. If R is a rotation in $\mathcal{O}(W)$, then R may be extended to a rotation in $\mathcal{O}(V)$ if we set $Rx = x$ for all $x \in W^{\perp}$ and extend by linearity. If a basis $\{x_1, x_2, x_3\}$ is chosen for V, with $x_1 \in W^{\perp}, x_2, x_3 \in W$, then the matrix representing R is

$$A = \begin{bmatrix} 1 & 0 & 0 \\ 0 & \cos\theta & -\sin\theta \\ 0 & \sin\theta & \cos\theta \end{bmatrix}.$$

By extending each transformation in a cyclic subgroup $\mathscr{C}_2^n$ of $\mathcal{O}(W)$ in this fashion, we obtain a cyclic subgroup of rotations in $\mathcal{O}(V)$, which will be denoted by $\mathscr{C}_3^n$.

On the other hand, if S is a reflection in $\mathcal{O}(W)$, then S may also be extended to a *rotation* in $\mathcal{O}(V)$—in fact to the rotation through the angle π having the reflecting line of S in W as its axis of rotation (see Exercise 2.4). More explicitly, define $Sx = -x$ for all $x \in W^{\perp}$, and extend by linearity. In this case, we may choose a basis $\{x_1, x_2, x_3\}$ for V with respect to which the matrix representing the extended transformation S is

$$A = \begin{bmatrix} -1 & 0 & 0 \\ 0 & \cos\theta & \sin\theta \\ 0 & \sin\theta & -\cos\theta \end{bmatrix}.$$

If each transformation T in a dihedral subgroup $\mathscr{H}_2^n$ of $\mathcal{O}(W)$ is extended to a rotation in $\mathcal{O}(V)$, as above, the resulting set of rotations is a subgroup of $\mathcal{O}(V)$ isomorphic with $\mathscr{H}_2^n$ (Exercise 2.9). This subgroup is also called a dihedral group and is denoted by $\mathscr{H}_3^n$. Observe that as subgroups of $\mathcal{O}(V)$ the groups of $\mathscr{C}_3^2$ and $\mathscr{H}_3^1$ each consist of the identity transformation and one rotation through the angle π, and so they are geometrically indistinguishable.

Since the finite subgroups of $\mathcal{O}(\mathscr{R}^2)$ are all symmetry groups of regular polygons, it is natural to consider next groups of rotations leaving invariant regular polyhedra in $\mathscr{R}^3$.

There are (up to similarity) only five regular (convex) polyhedra in $\mathscr{R}^3$—the tetrahedron, cube, octahedron, dodecahedron, and icosahedron (see Figure 2.5). They have been known since antiquity. The five regular solids, or Platonic solids, are discussed in Book XIII of Euclid's *Elements*, and it often has been suggested that the first twelve books of Euclid were intended only as an introduction to the regular solids (see [7], p. 13, or [26], p. 74). The first four solids were known to the Pythagoreans of the sixth century B.C., and probably much earlier. There is a story to the effect that Hippasus, one of the Pythagoreans, was shipwrecked and drowned (presumably by the gods) because he had claimed credit for the construction of a dodecahedron inscribed in a sphere, rather than crediting the discovery to Pythagoras as was customary. All five solids were systematically studied by Theatetus (circa 380 B.C.), and Euclid's account of them was based on the work of Theatetus.

It will be useful in this chapter, and for examples in the succeeding chapters, for the reader to have available paper or cardboard models of a tetrahedron, cube, and icosahedron. The tetrahedron and cube are easily constructed, but the icosahedron may be less familiar so we have included

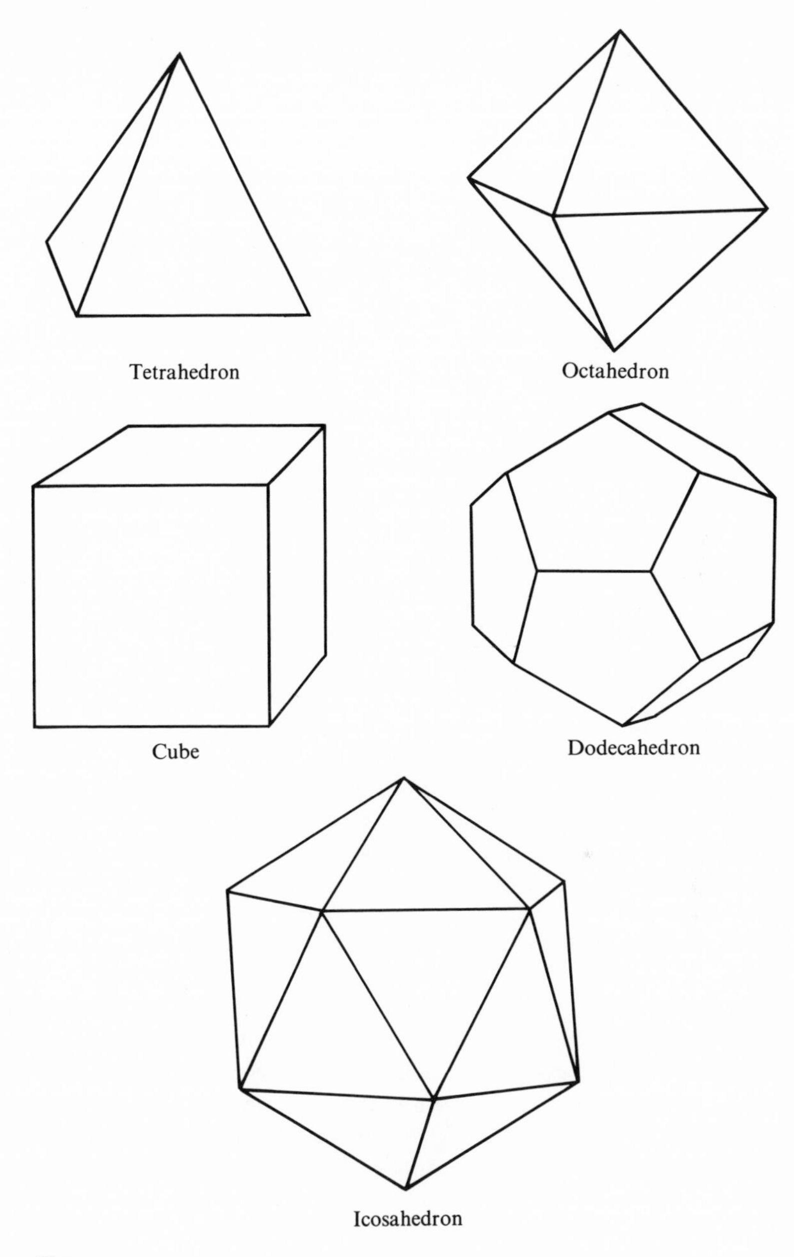

Figure 2.5

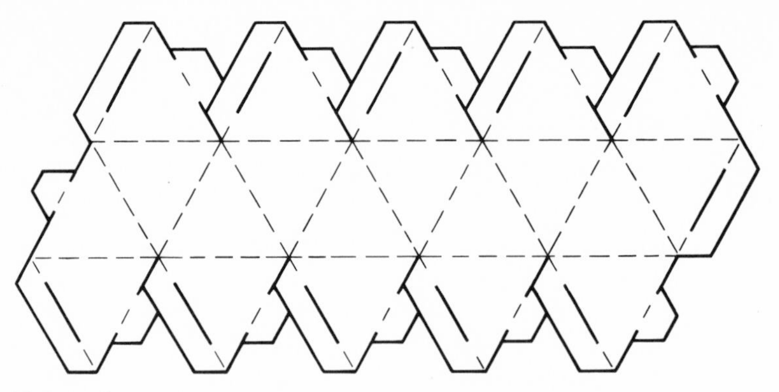

Figure 2.6

a sketch in Figure 2.6 showing how a model icosahedron can be constructed fairly easily from a single sheet. Solid lines indicate cuts and dashed lines indicate folds. More detailed instructions for the construction of models in general may be found in [10].

If a regular polyhedron is centered at the origin in $\mathscr{R}^3$, then the rotations in $\mathscr{O}(\mathscr{R}^3)$ that carry the polyhedron into itself constitute a finite subgroup of $\mathscr{O}(\mathscr{R}^3)$. Only three distinct finite groups of rotations arise in this manner, however. The cube has the same group of rotations as the octahedron, and the icosahedron has the same group as the dodecahedron. The reasons are geometrically very simple. If the midpoints of adjacent faces of a cube are joined by line segments, then the line segments are the edges of an octahedron inscribed in the cube. Any rotation of $\mathscr{R}^3$ that leaves the cube invariant also leaves the inscribed octahedron invariant, and vice versa. Similar remarks apply to the icosahedron and dodecahedron.

Let us discuss the rotation groups of the regular polyhedra in more detail.

Suppose that a tetrahedron is situated with its center at the origin in $\mathscr{R}^3$. The subgroup of rotations in $\mathscr{O}(\mathscr{R}^3)$ leaving the tetrahedron invariant will be denoted by $\mathscr{T}$. The elements of $\mathscr{T}$ consist of rotations through angles of $2\pi/3$ and $4\pi/3$ about each of four axes joining vertices with centers of opposite faces, rotations through the angle π about each of three axes joining the midpoints of opposite edges, and the identity. Thus

$$|\mathscr{T}| = 4 \cdot 2 + 3 \cdot 1 + 1 = 12.$$

The group of rotations of a cube centered at the origin will be denoted by $\mathscr{W}$. The elements of $\mathscr{W}$ are rotations of three distinct types, together

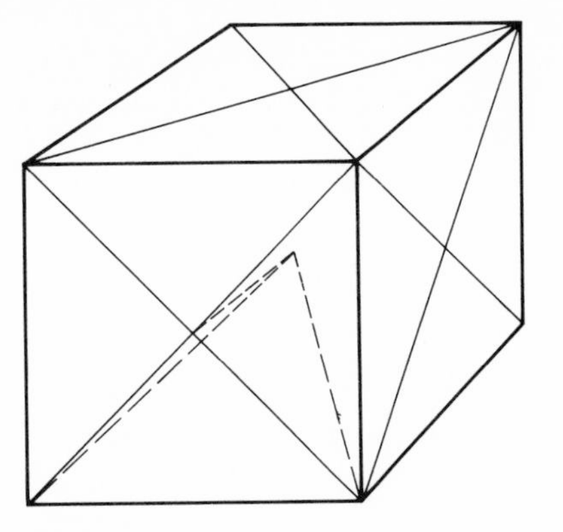

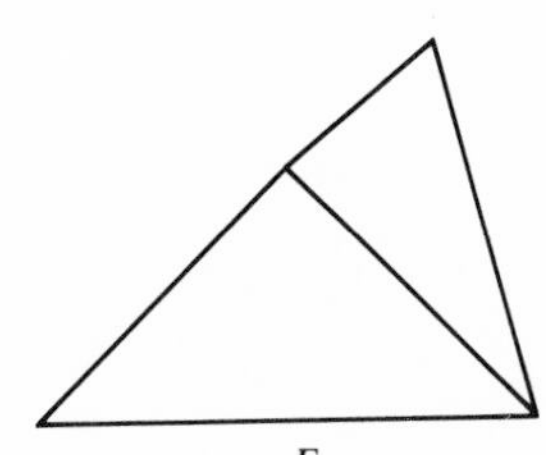

Figure 2.7

with the identity. There are rotations through angles of $\pi/2$, π, and $3\pi/2$ about each of three axes joining the centers of opposite faces, rotations through angles of $2\pi/3$ and $4\pi/3$ about each of four axes joining extreme opposite vertices, and rotations through the angle π about each of six axes joining midpoints of diagonally opposite edges. Thus

$$|\mathscr{W}| = 3 \cdot 3 + 4 \cdot 2 + 6 \cdot 1 + 1 = 24.$$

Denote by X the union of the six planes containing the diagonals of opposite faces of the cube. The complement of X in the cube has as connected components 24 congruent regions, each one a relatively open irregular tetrahedron (see Figure 2.7). Direct inspection shows that the component regions in the cube are permuted among themselves transitively by the elements of $\mathscr{W}$. Let F denote a fixed component. Since $\mathscr{W}$ acts transitively on the components we have

$$24 = |\mathrm{Orb}(F)| = [\mathscr{W} : \mathrm{Stab}(F)]$$

by Proposition 1.2.1. Thus $\mathrm{Stab}(F) = 1$, from which it follows that $F \cap RF = \varnothing$ if $1 \neq R \in \mathscr{W}$. Since $\bigcup \{(RF)^- : R \in \mathscr{W}\}$ is the entire cube, F is a fundamental region for $\mathscr{W}$ in the cube, in the sense discussed in Section 2.2.

The icosahedron has 20 faces, each one an equilateral triangle. It has 30 edges and 12 vertices. The rotation group $\mathscr{I}$ of the icosahedron consists of rotations through angles of $2\pi/5$, $4\pi/5$, $6\pi/5$, and $8\pi/5$ about each of 6 axes joining extreme opposite vertices, rotations through angles of $2\pi/3$ and $4\pi/3$ about each of 10 axes joining centers of opposite faces, rotations through the angle π about each of 15 axes joining midpoints of opposite edges, and the identity. Thus

$$|\mathscr{I}| = 6 \cdot 4 + 10 \cdot 2 + 15 \cdot 1 + 1 = 60.$$

The unit sphere $\{x \in V : \|x\| = 1\}$ is left invariant by every transformation $T \in \mathcal{O}(V)$. It is a consequence of Euler's theorem (2.3.1) that if $T \neq 1$ is a rotation, then there are precisely two points x on the unit sphere for which $Tx = x$: the points of intersection of the sphere and the axis of rotation for T. These two points will be called *poles* of T. If $\mathcal{G}$ is a subgroup of $\mathcal{O}(V)$, let us denote by $\mathcal{S}$ the set of poles of nonidentity rotations in $\mathcal{G}$.

Proposition 2.4.1

If $\dim V = 3$ and $\mathcal{G} \leq \mathcal{O}(V)$, then $\mathcal{G}$ is a permutation group on its set $\mathcal{S}$ of poles.

Proof

If $x \in \mathcal{S}$, then x is a pole for some rotation $T \in \mathcal{G}$, $T \neq 1$. For any $R \in \mathcal{G}$ we have

$$(RTR^{-1})Rx = RTx = Rx,$$

so Rx is a pole of the rotation RTR^{-1} and $Rx \in \mathcal{S}$.

Let us analyze the actions of the rotation groups discussed above as permutation groups on their sets of poles.

Each cyclic group $\mathcal{C}_3^n$ has exactly two poles. No rotation in $\mathcal{C}_3^n$ carries either pole to the other, so $\mathcal{S}$ has two one-element orbits, and the stabilizer of each pole has order n.

The dihedral group $\mathcal{H}_3^n$ has n axes of rotation in the plane W on which $\mathcal{H}_2^n$ acts, and one axis of rotation orthogonal to W, so $\mathcal{H}_3^n$ has $2n + 2$ poles. The two poles on the axis orthogonal to W constitute one orbit in $\mathcal{S}$. If n is odd, the n poles that are vertices of a regular n-gon in W constitute another orbit, and the set of n negatives of these points constitutes a third orbit. If n is even, the set of n vertices of the regular n-gon in W constitutes one orbit, and the n poles on the axes through the midpoints of opposite sides of the n-gon constitute a third orbit. Thus in each case $\mathcal{S}$ has three orbits, and the stabilizers of elements in each of the three orbits have orders n, 2, and 2, respectively.

The discussion of the three remaining groups, $\mathcal{T}$, $\mathcal{W}$, and $\mathcal{I}$, is left to the reader. We shall tabulate in Table 2.1 the groups, their orders, the number of poles, the number of orbits, and the orders of the stabilizers.

We shall show next that if $\mathcal{G}$ is any finite rotation subgroup of $\mathcal{O}(V)$, then the data involving the number of orbits of $\mathcal{S}$ and the orders of stabilizers must coincide with the data of one of the groups listed in Table 2.1.

$\mathscr{G}$	$\lvert\mathscr{G}\rvert$	Orbits	$\lvert\mathscr{S}\rvert$	Orders of stabilizers		
$\mathscr{C}_3^n$	n	2	2	n	n	
$\mathscr{H}_3^n$	$2n$	3	$2n+2$	2	2	n
$\mathscr{T}$	12	3	14	2	3	3
$\mathscr{W}$	24	3	26	2	3	4
$\mathscr{I}$	60	3	62	2	3	5

Table 2.1

Suppose then that $\mathscr{G}$ is a finite rotation group and denote by $\mathscr{U}$ the set of all ordered pairs (T, x), where $T \in \mathscr{G}$, $T \neq 1$, and $x \in \mathscr{S}$ is a pole of T. Let us denote $\lvert\mathscr{G}\rvert$ by n, and for each $x \in \mathscr{S}$ set

$$v_x = \lvert\text{Orb}(x)\rvert \qquad \text{and} \qquad n_x = \lvert\text{Stab}(x)\rvert.$$

Note that x lies on an axis of rotation and that n_x is simply the order of the cyclic subgroup of $\mathscr{G}$ each of whose elements is a rotation about that axis. By Proposition 1.2.1 we have $n = n_x v_x$ for each $x \in \mathscr{S}$.

Since each $T \neq 1$ in $\mathscr{G}$ has exactly two poles, we have $\lvert\mathscr{U}\rvert = 2(n-1)$. On the other hand, we may count the elements of $\mathscr{U}$ by counting the number of group elements corresponding to each pole. Suppose that $\{x_1, \ldots, x_k\}$ is a set of poles, one from each orbit in $\mathscr{S}$, and set $n_{x_i} = n_i$, $v_{x_i} = v_i$. Then since each $x \in \text{Orb}(x_i)$ has $n_x = n_i$, we have

$$\begin{aligned}\lvert\mathscr{U}\rvert &= \Sigma\{n_x - 1 : x \in \mathscr{S}\} \\ &= \Sigma_{i=1}^k v_i(n_i - 1) = \Sigma_{i=1}^k (n - v_i).\end{aligned}$$

Thus $2n - 2 = \Sigma_i(n - v_i)$, and dividing by n we have

$$2 - 2/n = \Sigma_{i=1}^k (1 - 1/n_i).$$

We may assume that $n > 1$; so $1 \leq 2 - 2/n < 2$. Since each $n_i \geq 2$, we have $1/2 \leq 1 - 1/n_i < 1$; so k must be either 2 or 3.

If $k = 2$, then

$$2 - 2/n = (1 - 1/n_1) + (1 - 1/n_2),$$

or $2 = n/n_1 + n/n_2 = v_1 + v_2$; so $v_1 = v_2 = 1$, $n_1 = n_2 = n$. Thus $\mathscr{G}$ has just one axis of rotation, and $\mathscr{G}$ is a cyclic group $\mathscr{C}_3^n$.

If $k = 3$, we may assume that $n_1 \leq n_2 \leq n_3$. If n_1 were 3 or greater, then we would have $\Sigma_i(1 - 1/n_i) \geq \Sigma_i(1 - 1/3) = 2$, a contradiction. Thus $n_1 = 2$ and we have

$$2 - 2/n = 1/2 + (1 - 1/n_2) + (1 - 1/n_3),$$

or

$$1/2 + 2/n = 1/n_2 + 1/n_3.$$

If n_2 were 4 or greater, we would have $1/n_2 + 1/n_3 \leq 1/2$, a contradiction; so $n_2 = 2$ or $n_2 = 3$.

If $n_2 = 2$, then $n_3 = n/2$, and we conclude that $v_1 = v_2 = n/2$, $v_3 = 2$. Setting $m = n/2$ we have obtained the data of the dihedral group $\mathscr{H}_3^m$.

If $n_2 = 3$, we have $1/6 + 2/n = 1/n_3$ and the only possibilities are

(a) $n_3 = 3, n = 12$,

(b) $n_3 = 4, n = 24$,

and

(c) $n_3 = 5, n = 60$,

since $n_3 \geq 6$ would require $2/n \leq 0$. When $n_3 = 3$, then $v_1 = 6$, $v_2 = v_3 = 4$ and we have the data of $\mathscr{T}$. When $n_3 = 4$, then $v_1 = 12$, $v_2 = 8$, $v_3 = 6$ and we have the data of $\mathscr{W}$. When $n_3 = 5$, then $v_1 = 30$, $v_2 = 20$, $v_3 = 12$, and the data are that of $\mathscr{I}$.

In each case the group $\mathscr{G}$ not only shares the data of Table 2.1 with one of the groups discussed earlier, but in fact *is* the group with that data, since the data are sufficient to determine the group. For example, when $n_1 = 2$, $n_2 = n_3 = 3$, and $n = 12$, the poles in either of the four-element orbits are the vertices of a regular tetrahedron centered at the origin. The tetrahedron is invariant under $\mathscr{G}$, so $\mathscr{G} \leq \mathscr{T}$; but also $|\mathscr{G}| = |\mathscr{T}| = 12$, so $\mathscr{G} = \mathscr{T}$. A more thorough discussion of this point for all the finite rotation groups may be found in [28], pp. 93–94.

We may conclude, finally, that

$$\mathscr{C}_3^n,\ n \geq 1;\ \mathscr{H}_3^n,\ n \geq 2;\ \mathscr{T};\ \mathscr{W};\ \text{and } \mathscr{I}$$

is a complete list of finite rotation subgroups of $\mathscr{O}(V)$ when $\dim V = 3$.

2.5 FINITE GROUPS IN THREE DIMENSIONS

The group $\mathscr{W}^*$ of all orthogonal transformations that leave a cube invariant is larger than the group $\mathscr{W}$ of rotations of the cube since, for example, -1 is an element of $\mathscr{W}^*$ but not of $\mathscr{W}$. Observe, however, that if $T \in \mathscr{W}^* \setminus \mathscr{W}$, then $-T = -1 \cdot T \in \mathscr{W}$ since $\det(-T) = 1$. Thus $\mathscr{W}^* = \mathscr{W} \cup (-1)\mathscr{W}$.

The above observations illustrate a general fact about subgroups of $\mathcal{O}(V)$, no matter what the dimension of V might be.

Proposition 2.5.1

If $\mathcal{G} \leq \mathcal{O}(V)$ and $\mathcal{H}$ is the rotation subgroup of $\mathcal{G}$, then either $\mathcal{H} = \mathcal{G}$ or else $[\mathcal{G} : \mathcal{H}] = 2$. In particular, $\mathcal{H}$ is a normal subgroup of $\mathcal{G}$.

Proof

Suppose that $T \in \mathcal{G} \setminus \mathcal{H}$. Then given any $S \in \mathcal{G} \setminus \mathcal{H}$, we have $\det(T^{-1}S) = (-1)^2 = 1$, so $T^{-1}S \in \mathcal{H}$; i.e., $S \in T\mathcal{H}$. Thus $\mathcal{G} = \mathcal{H} \cup T\mathcal{H}$ and $[\mathcal{G} : \mathcal{H}] = 2$.

Suppose again that $\dim V = 3$ and that $\mathcal{H}$ has index 2 in $\mathcal{G}$, and let us distinguish between two cases.

If $-1 \in \mathcal{G}$, then $\mathcal{G}$ is the union of $\mathcal{H}$ and the set of negatives of the transformations in $\mathcal{H}$. On the other hand, if $\mathcal{H}$ is any group of rotations in $\mathcal{O}(V)$, then $\mathcal{H} \cup \{-T : T \in \mathcal{H}\}$ is a subgroup of $\mathcal{O}(V)$ having $\mathcal{H}$ as its rotation subgroup [Exercise 2.13(a)].

Suppose then that $-1 \notin \mathcal{G}$, and that $R\mathcal{H}$ is the coset different from $\mathcal{H}$ in $\mathcal{G}$. Then $R^2 \in \mathcal{H}$, for $R^2 \in R\mathcal{H}$ would imply that $R \in \mathcal{H}$. Since $\mathcal{H}$ is normal in $\mathcal{G}$, we have $(-R\mathcal{H})(-R\mathcal{H}) = R^2\mathcal{H} = \mathcal{H}$ and $\mathcal{H}(-R\mathcal{H}) = -R\mathcal{H}$, from which it follows [Exercise 2.14(a)] that the set $\mathcal{K} = \mathcal{H} \cup (-R)\mathcal{H}$ is a group of rotations having $\mathcal{H}$ as a subgroup of index 2. Conversely, if $\mathcal{K}$ is any rotation group in $\mathcal{O}(V)$ having a subgroup $\mathcal{H}$ of index 2, then the set $\mathcal{G} = \mathcal{H} \cup \{-T : T \in \mathcal{K} \setminus \mathcal{H}\}$ is a subgroup of $\mathcal{O}(V)$ having $\mathcal{H}$ as its rotation subgroup [Exercise 2.14(b)].

We may now list all finite subgroups of $\mathcal{O}(V)$ for $\dim V = 3$. As we have seen, they divide naturally into three classes, the first class being groups of rotations. The second class consists of those groups obtained by choosing a group $\mathcal{H}$ of the first class and adjoining to it the negatives of all its elements. The resulting group will be denoted by $\mathcal{H}^*$. The third class consists of those groups obtained by choosing a group $\mathcal{K}$ of the first class having a subgroup $\mathcal{H}$ of index 2 and setting

$$\mathcal{G} = \mathcal{H} \cup \{-T : T \in \mathcal{K} \setminus \mathcal{H}\}.$$

We denote such a group by $\mathcal{K}]\mathcal{H}$. Note that $\mathcal{G}$ is of the second class if and only if $-1 \in \mathcal{G}$, and $\mathcal{G}$ is of the third class if and only if $\mathcal{G}$ is not of the first class but $-1 \notin \mathcal{G}$.

Using these facts and the list of rotation groups from Section 2.4, we list all finite subgroups of $\mathcal{O}(V)$ in the next theorem.

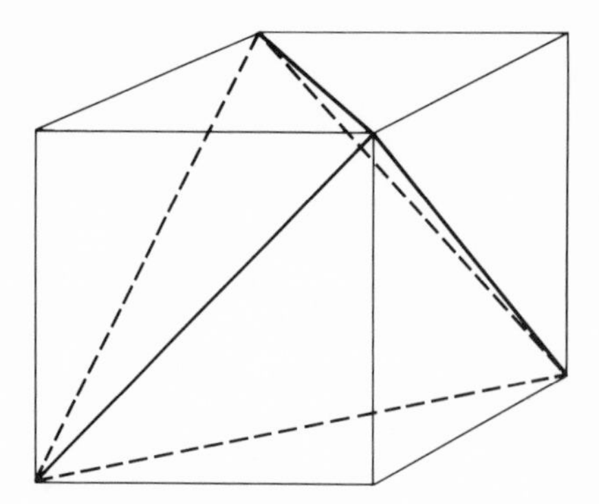

Figure 2.8

Theorem 2.5.2

If dim $V = 3$ and $\mathscr{G}$ is a finite subgroup of $\mathcal{O}(V)$, then $\mathscr{G}$ is one of the following:

(a) $\mathscr{C}_3^n, n \geq 1; \mathscr{H}_3^n, n \geq 2; \mathscr{T}; \mathscr{W}; \mathscr{I};$

(b) $(\mathscr{C}_3^n)^*, n \geq 1; (\mathscr{H}_3^n)^*, n \geq 2; \mathscr{T}^*; \mathscr{W}^*; \mathscr{I}^*;$

(c) $\mathscr{C}_3^{2n}]\mathscr{C}_3^n, n \geq 1; \mathscr{H}_3^n]\mathscr{C}_3^n, n \geq 2; \mathscr{H}_3^{2n}]\mathscr{H}_3^n, n \geq 2; \mathscr{W}]\mathscr{T}.$

Let us see geometrically why $\mathscr{T}$ is a subgroup of $\mathscr{W}$. As shown in Figure 2.8, a regular tetrahedron may be inscribed in a cube. Moreover, this tetrahedron is invariant under the rotations in $\mathscr{W}$ of order 3 about axes joining extreme opposite vertices, as well as rotations of order 2 about axes joining centers of opposite faces. These rotations, together with the identity, constitute the 12 elements of $\mathscr{T}$.

The groups $\mathscr{T}$ and $\mathscr{I}$ have no subgroups of index 2, as indicated in Exercises 2.16 and 2.17.

Observe that there are three groups of order 2—$\mathscr{C}_3^2$, $(\mathscr{C}_3^1)^*$, and $C_3^2]\mathscr{C}_3^1$—in the list of groups in Theorem 2.5.2. As abstract groups they are, of course, all isomorphic. However, they are geometrically different since their nonidentity elements are a rotation, an inversion through the origin, and a reflection, respectively. In order to see that the list in Theorem 2.5.2 is not redundant, let us give a precise definition of the phrase "geometrically the same." Two subgroups $\mathscr{G}_1$ and $\mathscr{G}_2$ of $\mathcal{O}(V)$ are considered to be geometrically the same if and only if $\mathscr{G}_2 = T\mathscr{G}_1T^{-1}$ for some $T \in \mathcal{O}(V)$ (see Exercises 2.7 and 2.8); otherwise, they are geometrically different.

With respect to suitable bases the nonidentity elements of $\mathscr{C}_3^2$, $(\mathscr{C}_3^1)^*$, and $\mathscr{C}_3^2]\mathscr{C}_3^1$ are represented by the matrices

$$\begin{bmatrix} 1 & 0 & 0 \\ 0 & -1 & 0 \\ 0 & 0 & -1 \end{bmatrix}, \quad \begin{bmatrix} -1 & 0 & 0 \\ 0 & -1 & 0 \\ 0 & 0 & -1 \end{bmatrix}, \quad \begin{bmatrix} 1 & 0 & 0 \\ 0 & 1 & 0 \\ 0 & 0 & -1 \end{bmatrix}.$$

Since the multiplicities of the eigenvalues are unequal, the groups $\mathscr{C}_3^2$, $(\mathscr{C}_3^1)^*$, and $\mathscr{C}_3^2]\mathscr{C}_3^1$ are geometrically different, according to the above definition. It is left to the reader to check that there are no redundancies among the remaining groups listed in Theorem 2.5.2.

2.6 *CRYSTALLOGRAPHIC GROUPS*

Suppose that $\dim V = 3$. A *lattice* in V is a discrete set of points obtained by taking all integer linear combinations of three linearly independent vectors x_1, x_2, and x_3. A subgroup $\mathscr{G}$ of $\mathcal{O}(V)$ is said to satisfy the *crystallographic condition*, or to be a *crystallographic* group, if and only if there is a lattice $\mathscr{L}$ invariant under $\mathscr{G}$; i.e., $Tx \in \mathscr{L}$ for all $T \in \mathscr{G}$, all $x \in \mathscr{L}$.

Suppose that T is a rotation in a crystallographic group $\mathscr{G}$. Then by Euler's theorem (2.3.1) there is a basis for V with respect to which T is represented by the matrix

$$A = \begin{bmatrix} 1 & 0 & 0 \\ 0 & \cos\theta & -\sin\theta \\ 0 & \sin\theta & \cos\theta \end{bmatrix},$$

where $0 \leq \theta < 2\pi$. Thus the trace of T is $1 + 2\cos\theta$.

On the other hand, we may choose as a basis for V the basic vectors x_1, x_2, and x_3 of a lattice $\mathscr{L}$ invariant under $\mathscr{G}$. Since each Tx_i is in $\mathscr{L}$ and is hence an integer linear combination of x_1, x_2, and x_3, the matrix representing T with respect to the basis $\{x_1, x_2, x_3\}$ has integer entries. Thus trace (T) is an integer, so $2\cos\theta$ is an integer, which is possible only if

$$\theta = 0, \pi/3, \pi/2, 2\pi/3, \pi, 4\pi/3, 3\pi/2, \text{ or } 5\pi/3.$$

It follows that either $T = 1$ or else T has order 2, 3, 4, or 6.

If $T \in \mathscr{G}$ is not a rotation, then $T = RS$, where S is a reflection and R is a rotation through an angle θ, $0 \leq \theta < 2\pi$, by Theorem 2.3.2 (and its proof). In this case trace $(T) = -1 + 2\cos\theta$ is an integer, so $2\cos\theta$ is an integer, and θ must be one of the angles listed above. If $R = 1$, then T

has order 2, and if R has order 3, then T has order 6. In all other cases R and T have the same order, so T must have order 2, 3, 4, or 6.

Among all finite subgroups of $\mathcal{O}(V)$ as listed in Theorem 2.5.2, only the following satisfy the above requirements on orders of elements:

(a) $\mathscr{C}_3^1, \mathscr{C}_3^2, \mathscr{C}_3^3, \mathscr{C}_3^4, \mathscr{C}_3^6, \mathscr{H}_3^2, \mathscr{H}_3^3, \mathscr{H}_3^4, \mathscr{H}_3^6, \mathscr{T}, \mathscr{W}$;

(b) $(\mathscr{C}_3^1)^*, (\mathscr{C}_3^2)^*, (\mathscr{C}_3^3)^*, (\mathscr{C}_3^4)^*, (\mathscr{C}_3^6)^*, (\mathscr{H}_3^2)^*, (\mathscr{H}_3^3)^*, (\mathscr{H}_3^4)^*, (\mathscr{H}_3^6)^*, \mathscr{T}^*, \mathscr{W}^*$;

(c) $\mathscr{C}_3^2]\mathscr{C}_3^1, \mathscr{C}_3^4]\mathscr{C}_3^2, \mathscr{C}_3^6]\mathscr{C}_3^3, \mathscr{H}_3^2]\mathscr{C}_3^2, \mathscr{H}_3^3]\mathscr{C}_3^3, \mathscr{H}_3^4]\mathscr{C}_3^4, \mathscr{H}_3^6]\mathscr{C}_3^6, \mathscr{H}_3^4]\mathscr{H}_3^2, \mathscr{H}_3^6]\mathscr{H}_3^3, \mathscr{W}]\mathscr{T}$.

It can be shown for each of the groups listed there is an invariant lattice (see [28], chap. 3; also see Exercise 2.20 and Section 5.2). Thus there are exactly 32 geometrically distinct finite crystallographic groups in three dimensions.

Exercises

2.1 Verify that

$$x_1 = (\cos\theta/2, \sin\theta/2),$$
$$x_2 = (-\sin\theta/2, \cos\theta/2)$$

are eigenvectors with respective eigenvalues 1 and -1 for the matrix

$$B = \begin{bmatrix} \cos\theta & \sin\theta \\ \sin\theta & -\cos\theta \end{bmatrix}.$$

2.2 Prove by induction that

$$\begin{bmatrix} \cos\theta & -\sin\theta \\ \sin\theta & \cos\theta \end{bmatrix}^m = \begin{bmatrix} \cos m\theta & -\sin m\theta \\ \sin m\theta & \cos m\theta \end{bmatrix}$$

for all positive integers m.

2.3 Define transformations S and R of $\mathscr{R}^2$ relative to the basis $\{e_1, e_2\}$ by the matrices

$$A = \begin{bmatrix} 1 & 0 \\ 0 & -1 \end{bmatrix} \quad \text{and} \quad B = \begin{bmatrix} \cos 2\pi/n & -\sin 2\pi/n \\ \sin 2\pi/n & \cos 2\pi/n \end{bmatrix},$$

respectively. Show that $S^2 = R^n = 1$ and that $RS = SR^{n-1}$. Conclude that $\langle R\rangle$ is the cyclic group $\mathscr{C}_2^n$ and that $\langle R, S\rangle$ is the dihedral group $\mathscr{H}_2^n$. Thus each of the groups $\mathscr{C}_2^n$ and $\mathscr{H}_2^n$ occurs as a subgroup of $\mathcal{O}(\mathscr{R}^2)$.

2.4 Suppose that dim $V = 3$ and that W is a plane in V. When a reflection in $\mathcal{O}(W)$ is extended to a rotation in $\mathcal{O}(V)$ as in Section 2.4, show that the axis of rotation is the original reflecting line in W.

2.5 Show that the product of two reflections of $\mathcal{R}^2$ is a rotation through twice the angle between their reflecting lines. More precisely, say that S_i has reflecting line at angle θ_i with the positive x-axis, with $0 \leq \theta_i < \pi$, and say that $\theta_1 \leq \theta_2$. Then $S_2 S_1$ is a counterclockwise rotation through angle $2(\theta_2 - \theta_1)$, and $S_1 S_2$ is a clockwise rotation through angle $2(\theta_2 - \theta_1)$.

2.6 Show that $\mathcal{W}]\mathcal{T}$ is the group of all symmetries of the tetrahedron.

2.7 If X and Y are regular n-gons of the same size both centered at the origin in $\mathcal{R}^2$, show that there is a transformation $T \in \mathcal{O}(\mathcal{R}^2)$ such that $TX = Y$. If $\mathcal{H}_1$ and $\mathcal{H}_2$ are the cyclic groups of rotations leaving X and Y invariant, respectively, show that $\mathcal{H}_2 = T\mathcal{H}_1 T^{-1}$. If $\mathcal{G}_1$ and $\mathcal{G}_2$ are the dihedral groups of all orthogonal transformations leaving X and Y invariant, show that $\mathcal{G}_2 = T\mathcal{G}_1 T^{-1}$. Conclude that any two cyclic (dihedral) groups of the same order in $\mathcal{O}(\mathcal{R}^2)$, and also in $\mathcal{O}(\mathcal{R}^3)$, are conjugate, and hence are geometrically the same.

2.8 Extend the scope of Exercise 2.7 to include the groups of the regular polyhedra in $\mathcal{R}^3$.

2.9 Show that $\mathcal{H}_2^n$ and $\mathcal{H}_3^n$ are isomorphic.

2.10 Find fundamental regions in the tetrahedron for the groups $\mathcal{T}$ and $\mathcal{W}]\mathcal{T}$.

2.11 Find fundamental regions in the icosahedron for the groups $\mathcal{I}$ and $\mathcal{I}^*$.

2.12 Give intuitive arguments supporting each of the following statements concerning regular (convex) polyhedra in a space V of dimension 3 (for a careful definition of "regular" see [7], pp. 15–16, or [8], p. 78).

(a) Each vertex of a regular polyhedron in V must be common to at least three faces.

(b) If $n \geq 6$ each interior angle of a regular n-gon is at least $2\pi/3$.

(c) The faces of a regular polyhedron must be equilateral triangles, squares, or regular pentagons.

(d) The interior angles of a regular pentagon are each $3\pi/5$, so at most three regular pentagons can share a vertex in a regular polyhedron.

(e) At most three squares can share a vertex in a regular polyhedron.

(f) At most five equilateral triangles can share a vertex in a regular polyhedron.

Conclude that there can be at most five distinct regular polyhedra in V.

2.13 Suppose that $\dim V$ is odd.

(a) If $\mathscr{H} \leq \mathcal{O}(V)$ is a rotation group, show that $\mathscr{H}^* = \mathscr{H} \cup \{-T: T \in \mathscr{H}\}$ is a subgroup of $\mathcal{O}(V)$ having $\mathscr{H}$ as its rotation subgroup.

(b) Suppose that $\mathscr{G} \leq \mathcal{O}(V)$ has rotation subgroup $\mathscr{H}$. Show that $\mathscr{H} \cup \{-T: T \in \mathscr{G} \setminus \mathscr{H}\}$ is a rotation group.

2.14 Suppose that $\dim V$ is odd.

(a) If $\mathscr{G} \leq \mathcal{O}(V)$ has rotation subgroup $\mathscr{H} \neq \mathscr{G}$, $-1 \notin \mathscr{G}$, and R is any element of $\mathscr{G} \setminus \mathscr{H}$, set $\mathscr{K} = \mathscr{H} \cup -R\mathscr{H}$. Show that $\mathscr{K}$ is a rotation subgroup of $\mathcal{O}(V)$ having $\mathscr{H}$ as a subgroup of index 2.

(b) If $\mathscr{K}$ is a rotation subgroup of $\mathcal{O}(V)$ having a subgroup $\mathscr{H}$ of index 2, show that

$$\mathscr{K}]\mathscr{H} = \mathscr{H} \cup \{-T : T \in \mathscr{K} \setminus \mathscr{H}\}$$

is a subgroup of $\mathcal{O}(V)$ having $\mathscr{H}$ as its rotation subgroup.

2.15 Show that $\mathscr{W}$ acts as a faithful permutation group on the set $\mathscr{S}$ of diagonals of the cube. Conclude that $\mathscr{W}$ is isomorphic with the symmetric group $\mathscr{S}_4$.

2.16 (a) By viewing $\mathscr{T}$ as a permutation group on the set $\mathscr{S}$ of vertices of the tetrahedron, show that $\mathscr{T}$ is isomorphic with the alternating group $\mathfrak{A}_4$ on four letters.

(b) Show that $\mathfrak{A}_4$ has two conjugacy classes of 3-cycles.

(c) If $\mathfrak{A}_4$ were to have a subgroup $\mathscr{H}$ of order 6, show that $\mathscr{H}$ must contain exactly four 3-cycles.

(d) Show that $\mathscr{H}$ must contain a permutation of the form $(ab)(cd)$ of order 2.

(e) Show also that

$$(ab)(cd) \neq (abc)^{-1}(ab)(cd)(abc) \in \mathscr{H},$$

contradicting $|\mathscr{H}| = 6$.

(f) Conclude that $\mathscr{T}$ has no subgroup of index 2.

2.17 Suppose that $\mathscr{H}$ is a normal subgroup of $\mathscr{I}$.

(a) Show that any two cyclic subgroups of $\mathscr{I}$ that have the same order are conjugate in $\mathscr{I}$.

(b) Show that $\mathscr{I}$ has 15 subgroups of order 2, 10 of order 3, and 6 of order 5.

(c) Show that $\mathscr{H}$ must contain either all or none of the cyclic subgroups of each of the orders 2, 3, and 5.

(d) Show that $|\mathscr{H}| = 1 + 15\alpha_1 + 20\alpha_2 + 24\alpha_3$, where each α_i is either 0 or 1.

(e) Use Lagrange's theorem to conclude that either $\alpha_i = 0$ for all i, or $\alpha_i = 1$ for all i. As a result $\mathscr{I}$ is a simple group; i.e., its only normal subgroups are 1 and $\mathscr{I}$.

(f) Conclude in particular that $\mathscr{I}$ has no subgroups of index 2.

2.18 There are 15 axes for rotations of order 2 joining midpoints of opposite edges of the icosahedron. If l_1 is one such axis, there are two others, l_2 and l_3, that are perpendicular to l_1 and to one another.

(a) Show that the mutually perpendicular axes l_1, l_2, and l_3 are the axes of rotation for $\mathscr{H}_3^2$, and hence that $\mathscr{H}_3^2$ occurs five times as a subgroup of $\mathscr{I}$.

(b) Let $\mathscr{S}$ be the set of five triples (l_1, l_2, l_3) of mutually perpendicular axes of order 2. Show that $\mathscr{I}$ acts as a permutation group on $\mathscr{S}$.

(c) Show that $\mathscr{I}$ is isomorphic with the alternating group on five letters.

2.19 Show that if $\mathscr{G}_1$ and $\mathscr{G}_2$ are two distinct groups in the list of Theorem 2.5.2, then $\mathscr{G}_1$ and $\mathscr{G}_2$ are geometrically different.

2.20 (a) Suppose that the matrices

$$A = \begin{bmatrix} 1 & 0 & 0 \\ 0 & 0 & -1 \\ 0 & 1 & 0 \end{bmatrix} \quad \text{and} \quad B = \begin{bmatrix} -1 & 0 & 0 \\ 0 & 1 & 0 \\ 0 & 0 & 1 \end{bmatrix}$$

represent a generating rotation and reflection for $\mathscr{H}_3^4$ in $\mathscr{O}(\mathscr{R}^3)$. Let $\mathscr{L}$ be the lattice $\{\Sigma_{i=1}^3 n_i e_i : n_i \in \mathbb{Z}\}$. Show that $\mathscr{L}$ is invariant under $\mathscr{H}_3^4$.

(b) Find invariant lattices for more of the crystallographic groups.

2.21 Determine which of the groups discussed in this chapter are generated by the reflections they contain.

2.22 If γ is the complex number $\cos 2\pi/5 + i \sin 2\pi/5$, then $\gamma^5 = 1$ by DeMoivre's theorem, so γ is a root of the polynomial $x^5 - 1$. Since

$$x^5 - 1 = (x - 1)(x^4 + x^3 + x^2 + x + 1)$$

and $\gamma \neq 1$, γ is in fact a root of

$$f(x) = x^4 + x^3 + x^2 + x + 1 = x^2\left(x^2 + x + 1 + \frac{1}{x} + \frac{1}{x^2}\right).$$

(a) Make the substitution $y = x + 1/x$ within the parentheses and find explicitly all four roots of $f(x)$.
(b) Set $\alpha = \cos \pi/5$ and $\beta = \cos 2\pi/5$. Show that $\beta = (-1 + \sqrt{5})/4$ and use this to show that $\alpha = (1 + \sqrt{5})/4$.
(c) Show that $4\alpha^2 = 2\alpha + 1$, $4\beta^2 = -2\beta + 1$, $2\alpha = 2\beta + 1$, and $4\alpha\beta = 1$.

2.23 A simple geometrical construction of an icosahedron is given in [13], vol. 3, pp. 491–492.
(a) If the edge length of the cube is taken to be $4\alpha = 4\cos \pi/5$, show that each edge of the constructed icosahedron has length $a = 2$.
(b) If the line segments OM, ON, and OL are taken to lie on the coordinate axes in $\mathscr{R}^3$, show that the vertices of the icosahedron are the points

$$(\pm 1, 0, \pm 2\alpha), \qquad (\pm 2\alpha, \pm 1, 0), \qquad (0, \pm 2\alpha, \pm 1).$$

2.24 A line segment is divided by the *golden section* if the ratio of the shorter to the longer section equals the ratio of the longer to the whole segment. If the shorter section has length 1 and the longer has length τ, show that $\tau = 2\alpha = 2\cos \pi/5$ (see Exercise 2.22).

2.25 The *Fibonacci numbers* are defined recursively as follows:

$$\varphi_0 = 0, \quad \varphi_1 = 1, \quad \varphi_n = \varphi_{n-1} + \varphi_{n-2}, \qquad \text{all } n \geq 2.$$

If τ is the number defined in Exercise 2.24, show that

$$\tau^n = \varphi_n \tau + \varphi_{n-1}$$

for all $n \geq 2$ [see Exercise 2.22(c) and use induction].

chapter 3

FUNDAMENTAL REGIONS

In Chapter 2 we met the notion of a fundamental region for certain finite subgroups of $\mathcal{O}(V)$, where dim V was either 2 or 3. We now present a formal definition in a more general setting.

Suppose that $\mathcal{G}$ is a finite subgroup of $\mathcal{O}(V)$. A subset F of V is called a *fundamental region* for $\mathcal{G}$ in V if and only if

(1) F is open,
(2) $F \cap TF = \varnothing$ if $1 \neq T \in \mathcal{G}$,

and

(3) $V = \bigcup\{(TF)^- : T \in \mathcal{G}\}$.

More generally, if X is a subset of V invariant under $\mathcal{G}$, then a subset F of X is a *fundamental region* for $\mathcal{G}$ in X if and only if

(1) F is relatively open in X,
(2) $F \cap TF = \varnothing$ if $1 \neq T \in \mathcal{G}$,

and

(3) $X = \bigcup\{(TF)^- \cap X : T \in \mathcal{G}\}$.

The purpose of this chapter is to describe a construction that yields a fundamental region for any finite subgroup of $\mathcal{O}(V)$. The construction was utilized by Fricke and Klein in the study of automorphic functions (see [15], p. 108).

Proposition 3.1.1

If $\dim V \geq 1$, then V is not the union of a finite number of proper subspaces.

Proof

If $\dim V = 1$, then 0 is the only proper subspace of V. Assume that the proposition holds for spaces of dimension $n - 1$, where $n \geq 2$ is the dimension of V. Suppose that $V = V_1 \cup \cdots \cup V_m$ with each V_i a proper subspace, and let W be any subspace of V of dimension $n - 1$. Then

$$W = W \cap V = W \cap (\cup V_i) = (W \cap V_1) \cup \cdots \cup (W \cap V_m).$$

By the induction hypothesis $W = W \cap V_i$ for some i. Since $\dim W = n - 1$, $\dim V_i \leq n - 1$, and $W \subseteq V_i$, we may conclude that $W = V_i$. We have shown that every subspace W of dimension $n - 1$ occurs as one of the subspaces $V_1, \ldots, V_m$. This is a contradiction since V has infinitely many subspaces of dimension $n - 1$ (see Exercise 3.1).

Suppose now that $\mathcal{G} \neq 1$ is a finite subgroup of $\mathcal{O}(V)$. Since each $T \in \mathcal{G}$ is a linear transformation, it is immediate that the set

$$V_T = \{x \in V : Tx = x\}$$

is a subspace of V, since V_T is the null space of $T - 1$. If $T \neq 1$, then V_T is a proper subspace. By Proposition 3.1.1

$$V \neq \cup \{V_T : 1 \neq T \in \mathcal{G}\},$$

so we may choose a point $x_0 \in V$ that is not left fixed by any nonidentity element of $\mathcal{G}$. In the language of permutation groups $\mathrm{Stab}(x_0) = 1$, so

$$|\mathrm{Orb}(x_0)| = [\mathcal{G} : 1] = |\mathcal{G}|$$

by Proposition 1.2.1.

If $|\mathcal{G}| = N$, let us label its elements as $T_0, T_1, \ldots, T_{N-1}$ with $T_0 = 1$, and set $x_i = T_i x_0$, $0 \leq i \leq N - 1$, so that

$$\mathrm{Orb}(x_0) = \{x_0, x_1, \ldots, x_{N-1}\}.$$

If $i \neq 0$, the line segment $[x_0 x_i]$ is defined by

$$[x_0 x_i] = \{x_0 + \lambda(x_i - x_0) : 0 \leq \lambda \leq 1\},$$

so $x_i - x_0$ is a vector parallel to $[x_0 x_i]$. The midpoint of $[x_0 x_i]$ is the vector $(1/2)(x_0 + x_i)$, since

$$d(x_0, (1/2)(x_0 + x_i)) = d(x_i, (1/2)(x_0 + x_i)).$$

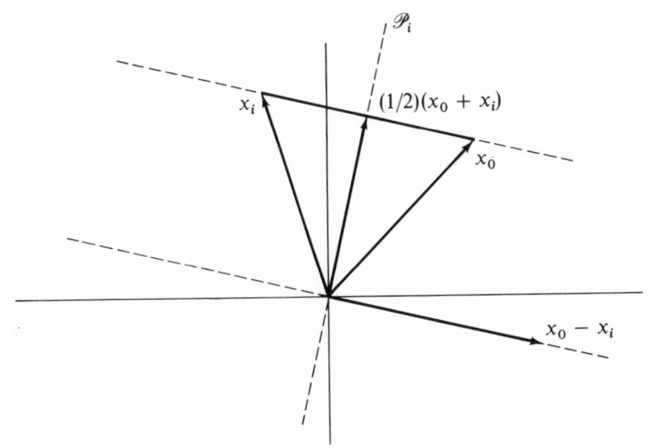

Figure 3.1

If the hyperplane $(x_0 - x_i)^\perp$ is denoted by $\mathscr{P}_i$, then the midpoint $(1/2)(x_0 + x_i)$ is in $\mathscr{P}_i$, since

$$((1/2)(x_0 + x_i), x_0 - x_i) = \frac{\|x_0\|^2 - \|x_i\|^2}{2}$$
$$= \frac{\|x_0\|^2 - \|T_i x_0\|^2}{2} = \frac{\|x_0\|^2 - \|x_0\|^2}{2} = 0.$$

Thus $\mathscr{P}_i$ is the "perpendicular bisector" of the line segment $[x_0 x_i]$ (see Figure 3.1).

If $x \in V$, it is easy to see (Exercise 3.4) that $x \perp (x_0 - x_i)$ if and only if $d(x, x_0) = d(x, x_i)$, using the fact that $x_i = T_i x_0$. It follows that

$$\mathscr{P}_i = \{x \in V : d(x, x_0) = d(x, x_i)\},$$

as might be expected of a perpendicular bisector.

Denote by L_i the open set

$$\{x \in V : d(x, x_0) < d(x, x_i)\},$$

$1 \le i \le N - 1$. The set L_i is called a *half-space determined by* $\mathscr{P}_i$ and can be thought of as the set of all points that are on the same side of the hyperplane $\mathscr{P}_i$ as x_0 is. Set $F = \bigcap \{L_i : 1 \le i \le N - 1\}$.

Theorem 3.1.2

The set F is a fundamental region for $\mathscr{G}$ in V.

Proof
Since each L_i is open, F is open. If $T_i \neq 1$, then $T_iF = T_i(\bigcap L_j)$, or

$$\begin{aligned} T_iF &= T_i(\{x : d(x, x_0) < d(x, x_j), 1 \leq j \leq N-1\}) \\ &= \{T_ix : d(T_ix, T_ix_0) < d(T_ix, T_iT_jx_0), 1 \leq i \leq N-1\} \\ &= \{y : d(y, x_i) < d(y, T_kx_0), 0 \leq k \leq N-1, k \neq i\}, \end{aligned}$$

since $\{T_iT_j : 1 \leq j \leq N-1\} = \mathcal{G} \setminus \{T_i\}$. Thus

$$T_iF = \{x : d(x, x_i) < d(x, x_j), \text{all } j \neq i\}.$$

If $x \in F \cap T_iF$, then $d(x, x_0) < d(x, x_i)$, and also $d(x, x_i) < d(x, x_0)$, a contradiction; so $F \cap T_iF = \varnothing$ for all $T_i \neq 1$. Finally, if $x \in V$, choose an index i for which $d(x, x_i)$ is minimal, and hence $d(x, x_i) \leq d(x, x_j)$ for all j. Since

$$(T_iF)^- = \{x : d(x, x_i) \leq d(x, x_j), 0 \leq j \leq N-1\},$$

we have $x \in (T_iF)^-$ (see Exercise 3.6). Thus

$$V = \bigcup \{(T_iF)^- : 0 \leq i \leq N-1\}$$

and F is a fundamental region.

Observe that the fundamental region F in Theorem 3.1.2 is connected, and in fact convex, being the intersection of convex sets.

The procedure indicated in Theorem 3.1.2 can be used to construct a fundamental region F_X for $\mathcal{G}$ in any set $X \subseteq V$ that is invariant under the action of $\mathcal{G}$, provided that it is possible to choose the point x_0 in X. In that case we simply define F_X to be $F \cap X$, and it is easily checked that F_X is a fundamental region for $\mathcal{G}$ in X (see Exercise 3.7).

Let us illustrate by constructing a fundamental region in the cube for the group $\mathcal{W}$ of rotations of the cube.

Divide each of the faces of the cube into four congruent squares, and let x_0 be the point at the center of one of the smaller squares. Then the various rotations of $\mathcal{W}$ carry x_0 to each of the centers of the 24 smaller squares on the surface of the cube (see Figure 3.2). If the transformations in $\mathcal{W}$ are labeled so that $T_1, \ldots, T_5$ carry x_0 to the points $x_1, \ldots, x_5$ as indicated in Figure 3.2, we may first intersect the half-spaces L_1, L_2, and L_3 to obtain the smaller shaded cube. If the smaller cube is then intersected with the half-spaces L_4 and L_5, it is easy to see that the remaining half-spaces are superfluous for the determination of their intersection, and that the resulting irregular pyramid F_X (Figure 3.3) is a fundamental region for $\mathcal{W}$ in the cube.

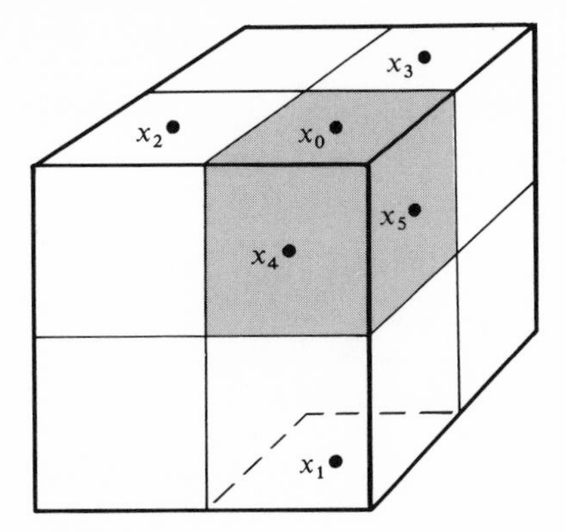

Figure 3.2

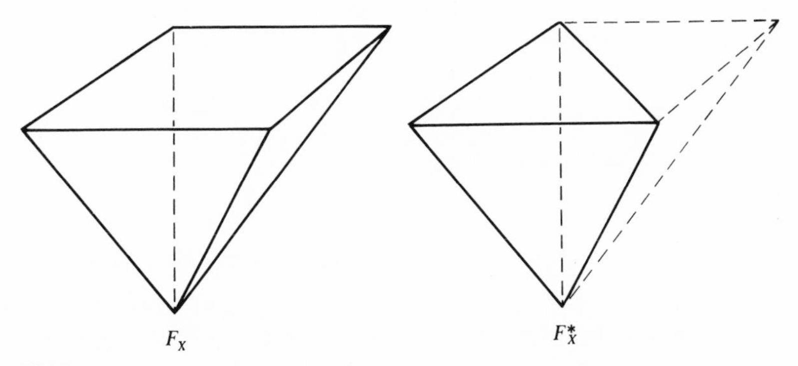

Figure 3.3

Observe that the fundamental region just constructed for $\mathscr{W}$ is different from the one presented in Chapter 2. The fundamental region of Chapter 2 can be obtained by the same procedure, however, if the point x_0 is chosen near the midpoint of an edge of one of the faces of the cube, halfway between the two adjacent edges.

Neither of the above choices for x_0 is suitable if we consider $\mathscr{G} = \mathscr{W}^*$, the group of all symmetries of the cube, since there are reflections in $\mathscr{W}^*$ leaving those points fixed. If we choose x_0 in the interior of one of the smaller squares on a face of the cube but off the diagonals of that smaller square, then it is not difficult to see that the fundamental region F_X^* obtained for $\mathscr{W}^*$ in the cube is just one half of the region F_X (see Figure 3.3), an irregular tetrahedron.

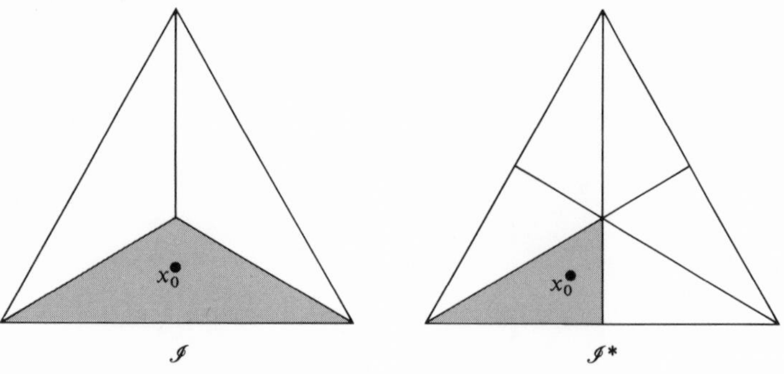

Figure 3.4

It is illuminating to construct the fundamental regions F_X and F_X^* in a modeling clay cube. The intersections of the cube with the half-spaces L_i are easily obtained by cutting the cube with a knife along the planes $\mathscr{P}_i$.

If X is the surface of an icosahedron, then the Fricke–Klein construction may be applied to obtain fundamental regions in X for $\mathscr{I}$ and $\mathscr{I}^*$ on a face of the icosahedron as indicated in Figure 3.4.

Exercises

3.1 If $\dim V \geq 2$, choose linearly independent vectors x_1 and x_2 in V. For each $\lambda \in \mathscr{R}$ define $V_\lambda = (x_1 + \lambda x_2)^\perp$. If $\lambda \neq \mu$, show that V_λ and V_μ are distinct $(n-1)$-dimensional subspaces of V.

3.2 If V is a vector space of dimension 2 or greater over any infinite field, show that V is not the union of a finite number of proper subspaces.

3.3 Show that the conclusion of Exercise 3.2 may fail if V is a vector space over a finite field.

3.4 Suppose that $x, y \in V$, $T \in \mathscr{O}(V)$, and $z = Ty$. Show that $x \perp (y - z)$ if and only if $d(x, y) = d(x, z)$.

3.5 If $F \subseteq V$ is a fundamental region for a group $\mathscr{G} \leq \mathscr{O}(V)$ and $T \in \mathscr{G}$, show that TF is also a fundamental region for $\mathscr{G}$ in V.

3.6 If x_i, $\mathscr{P}_i$, and L_i are as in the proof of Theorem 3.1.2, show that $L_i^- = L_i \cup \mathscr{P}_i$; so

$$L_i^- = \{x \in V : d(x, x_0) \leq d(x, x_i)\}.$$

Conclude that

$$F^- = \{x : d(x, x_0) \leq d(x, x_i), 0 \leq i \leq N - 1\},$$

and, more generally, that

$$(T_iF)^- = \{x : d(x, x_i) \leq d(x, x_j), 0 \leq j \leq N - 1\}.$$

3.7 If F is a fundamental region for $\mathcal{G}$ in V and $X \subseteq V$ is invariant under $\mathcal{G}$, show that $F_X = F \cap X$ is a fundamental region for $\mathcal{G}$ in X.

3.8 If Y is any finite subset of V with $0 \notin Y$, use Proposition 3.1.1 to show that there is a vector $t \in V$ such that $(y, t) \neq 0$ for all $y \in Y$.

3.9 Prove that an open half-space L determined by a hyperplane $\mathcal{P}$ is convex.

3.10 Construct fundamental regions in a regular n-gon in $\mathcal{R}^2$ for the groups $\mathcal{C}_2^n$ and $\mathcal{H}_2^n$.

3.11 Construct fundamental regions in a *dodecahedron* for the groups $\mathcal{I}$ and $\mathcal{I}^*$.

3.12 Construct fundamental regions in a tetrahedron for the groups $\mathcal{T}$ and $\mathcal{W}]\mathcal{T}$.

chapter 4

COXETER GROUPS

4.1 ROOT SYSTEMS

As we saw in Chapter 2, the description of all finite subgroups of $\mathcal{O}(\mathcal{R}^3)$ is rather involved. The enumeration of finite subgroups of $\mathcal{O}(V)$ becomes considerably more involved when V has dimension greater than 3. When $n = 4$ the subgroups are discussed in [2], and when $n \geq 5$ no complete listing is known. Thus we limit our discussion at this point to the primary subject of the book—the important class of finite subgroups of $\mathcal{O}(V)$ that are generated by reflections.

A *reflection* of V is a linear transformation S that carries each vector to its mirror image with respect to a fixed hyperplane $\mathscr{P}$. More precisely, $Sx = x$ if $x \in \mathscr{P}$ and $Sx = -x$ if $x \in \mathscr{P}^\perp$. Suppose that $0 \neq r \in \mathscr{P}^\perp$. If we define a transformation S_r by setting

$$S_r x = x - \frac{2(x, r)r}{(r, r)}$$

for all $x \in V$, then $S_r x = x$ if $x \in \mathscr{P}$, and $S_r r = r - 2r = -r$. Since $\mathscr{P} \cup \{r\}$ contains a basis for V, it follows that S_r is the reflection S. We shall speak of S_r as being the reflection *through* $\mathscr{P}$ or the reflection *along* r. Observe that $S_r = S_{\lambda r}$ for all $\lambda \neq 0$ and that $S_r^2 = 1$. It is clear geometrically, and it also follows easily from the formula defining S_r, that S_r is orthogonal (see Exercise 4.1).

Suppose that $\mathscr{G} \leq \mathcal{O}(V)$ and that $S \in \mathscr{G}$ is a reflection through a hyperplane $\mathscr{P}$. The two unit vectors $\pm r$ that are perpendicular to $\mathscr{P}$, so that $S = S_r$, are called *roots* of $\mathscr{G}$. Since $S_r = S_{\lambda r}$ for all $\lambda \neq 0$, the singling

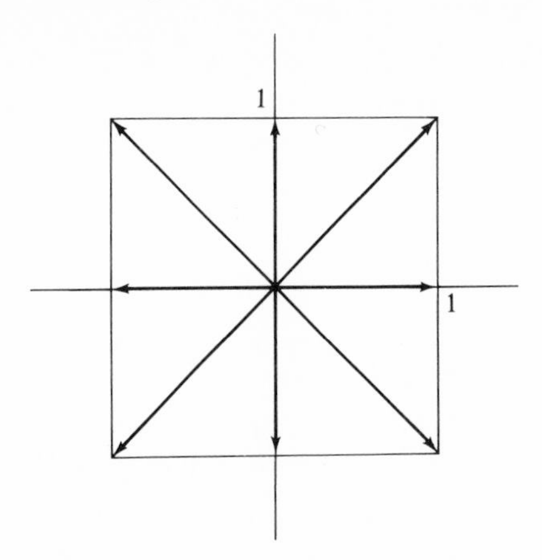

Figure 4.1

out of *unit* vectors as roots of $\mathscr{G}$ may seem somewhat arbitrary. In fact, for particular groups $\mathscr{G}$ there are choices of vectors r determining the reflections of $\mathscr{G}$ that seem more natural. For example, if the dihedral group $\mathscr{H}_2^4$ is viewed as the symmetry group of the square with vertices $(\pm 1, \pm 1)$ in $\mathscr{R}^2$ (see Figure 4.1), it is perhaps "natural" to choose the vectors

$$\{(\pm 1, 0), (0, \pm 1), (\pm 1, \pm 1)\}$$

as the set of roots of $\mathscr{H}_2^4$.

As we shall see in Chapter 5, there are analogous "natural" choices of relative lengths of roots for many reflection groups. Thus our choice of unit vectors as roots may be viewed as a temporary expediency. The results of this chapter (and their proofs) do not depend in an essential way on the relative lengths of roots, so the roots are taken to be unit vectors for the sake of convenience. For particular illustrative examples, such as $\mathscr{H}_2^4$, we shall feel free to assign other lengths to roots.

Proposition 4.1.1

If r is a root of $\mathscr{G} \leq \mathscr{O}(V)$ and if $T \in \mathscr{G}$, then Tr is also a root of $\mathscr{G}$. In fact, if $Tr = x$, then $S_x = TS_rT^{-1} \in \mathscr{G}$.

Proof

Set $\mathscr{P} = r^\perp$ and $\mathscr{P}' = T\mathscr{P}$. Then $\mathscr{P}'$ is a hyperplane, and

$$\mathscr{P}' = (Tr)^\perp = x^\perp$$

since $T \in \mathscr{O}(V)$. If $y = Tz \in \mathscr{P}'$, with $z \in \mathscr{P}$, then

$$TS_rT^{-1}y = TS_rz = Tz = y.$$

Also

$$TS_rT^{-1}x = TS_r r = -Tr = -x,$$

so $S_x = TS_rT^{-1} \in \mathscr{G}$.

If $\mathscr{G}$ is any subgroup of $\mathcal{O}(V)$, set

$$V_0 = V_0(\mathscr{G}) = \bigcap \{V_T : T \in \mathscr{G}\},$$

where V_T is the subspace $\{x \in V : Tx = x\}$. Then V_0 is a subspace of V, $T|V_0$ is the identity transformation on V_0 for every $T \in \mathscr{G}$, and V_0 is the largest subspace of V with that property. In particular, $TV_0 = V_0$ for all $T \in \mathscr{G}$; so $T(V_0^\perp) = V_0^\perp$ for all $T \in \mathscr{G}$. If V is represented as $V_0 \oplus V_0^\perp$, then every $T \in \mathscr{G}$ can be represented as $1 \oplus T'$, where $T' = T|V_0^\perp$. The group

$$\mathscr{G}' = \{T' : T \in \mathscr{G}\} \leq \mathcal{O}(V_0^\perp)$$

is clearly isomorphic with $\mathscr{G}$, and $V_0(\mathscr{G}') = 0$. Since the transformations in $\mathscr{G}'$ extend to those in $\mathscr{G}$ in a geometrically trivial fashion there is no loss of generality in studying only groups $\mathscr{G}$ for which $V_0(\mathscr{G}) = 0$. A subgroup $\mathscr{G}$ of $\mathcal{O}(V)$ with $V_0(\mathscr{G}) = 0$ will be called *effective*.

Proposition 4.1.2

Suppose that $\mathscr{G} \leq \mathcal{O}(V)$ is generated by reflections along roots $r_1, \ldots, r_k$. Then $\mathscr{G}$ is effective if and only if $\{r_1, \ldots, r_k\}$ contains a basis for V.

Proof

Set $W = \bigcap \{r_i^\perp : 1 \leq i \leq k\}$. Since the reflection along r_i acts as the identity transformation on $r_i^\perp$ and each $T \in \mathscr{G}$ is a product of the generating reflections, we have $T|W = 1_W$ for all $T \in \mathscr{G}$. Thus $W \subseteq V_0(\mathscr{G})$. On the other hand, if $x \in V_0$, then, in particular, each generating reflection leaves x invariant, so $x \in r_i^\perp$ for each i. Thus $x \in W$, and $W = V_0(\mathscr{G})$. Consequently, $\mathscr{G}$ is effective if and only if $W = 0$, or $W^\perp = V$. But

$$W^\perp = (\textstyle\bigcap_{i=1}^k r_i^\perp)^\perp = \Sigma_{i=1}^k r_i^{\perp\perp}.$$

In other words, the set $\{r_1, \ldots, r_k\}$ spans $W^\perp$, since $r_i^{\perp\perp}$ is the subspace spanned by r_i. Thus $\mathscr{G}$ is effective if and only if $\{r_1, \ldots, r_k\}$ spans V.

Suppose that $\mathscr{G} \leq \mathcal{O}(V)$ is generated by a finite set of reflections. We shall denote by Δ the set of all roots corresponding to the generating reflections, together with all images of these roots under all transformations in $\mathscr{G}$. Equivalently, by Proposition 4.1.1, Δ is the set of all roots corresponding to the reflections TST^{-1}, where T ranges over $\mathscr{G}$ and S ranges over the generating set of reflections. The set Δ will be called a *root system* for $\mathscr{G}$.

It appears possible at this point that the root system Δ of $\mathscr{G}$ is dependent on the particular generating set of reflections. It will be shown later (Theorem 4.2.4) that if $\mathscr{G}$ is finite, then Δ is the set of *all* roots of $\mathscr{G}$. In view of this fact it would perhaps seem more reasonable simply to define Δ to be the set of all roots of $\mathscr{G}$. There are, however, technical reasons involving the construction of reflection groups (in Section 5.3) for pursuing the present course.

The terminology of "roots" and "root systems" derives from the study of Lie algebras. In that context relative lengths of roots play an important role, and the notion of root system is more restrictive, in that the root systems of Lie algebras are required to satisfy a crystallographic condition (see Sections 2.6 and 5.2).

Proposition 4.1.3

Suppose that $\mathscr{G} \leq \mathcal{O}(V)$ is generated by a finite set of reflections, and that $\mathscr{G}$ is effective. If the root system Δ is finite, then $\mathscr{G}$ is finite.

Proof

By the definition of root system we have $T(\Delta) = \Delta$ for all $T \in \mathscr{G}$. Thus by restricting each $T \in \mathscr{G}$ to Δ we may view $\mathscr{G}$ as a permutation group on Δ. Since $\mathscr{G}$ is effective, Δ contains a basis for V by Proposition 4.1.2; so if $T|\Delta$ is the identity map on Δ, then $T = 1$. But that means that $\mathscr{G}$ is faithful on Δ, so $\mathscr{G}$ is finite if Δ is finite.

A finite effective subgroup $\mathscr{G}$ of $\mathcal{O}(V)$ that is generated by a set of reflections will be called a *Coxeter group*. For example, the dihedral groups $\mathscr{H}_2^n$, $n \geq 1$, are Coxeter groups, as are $\mathscr{W}^*$, $\mathscr{W}$] $\mathscr{T}$, and $\mathscr{I}^*$.

We shall assume for the remainder of this chapter that $\mathscr{G}$ is a Coxeter group, with root system Δ.

Choose a vector $t \in V$ such that $(t, r) \neq 0$ for every root r of $\mathscr{G}$ (see Exercise 3.8). Then the root system Δ is partitioned into two subsets,

$$\Delta_t^+ = \{r \in \Delta : (t, r) > 0\},$$

and

$$\Delta_t^- = \{r \in \Delta : (t, r) < 0\}.$$

Geometrically, Δ_t^+ and Δ_t^- are the subsets of Δ lying on the two sides of the hyperplane $t^\perp$. If $r \in \Delta$, then also $-r \in \Delta$, by definition, and $(t, -r) = -(t, r)$. Thus $r \in \Delta_t^+$ if and only if $-r \in \Delta_t^-$, and so $|\Delta_t^+| = |\Delta_t^-|$.

Choose a subset Π of Δ_t^+ that is minimal with respect to the property that every $r \in \Delta_t^+$ is a linear combination, with all coefficients nonnegative, of elements of Π. In other words, if Γ is any proper subset of Π, then there is a root $r \in \Delta_t^+$ that cannot be written as a nonnegative linear combination of elements of Γ. Such a minimal subset Π will be called a *t-base* for Δ. On the surface it is conceivable that $\Pi = \Delta_t^+$, but at any rate it is clear that at least one t-base exists, since Δ is a finite set. Since $\Delta_t^- = -\Delta_t^+$, every $r \in \Delta_t^-$ is a linear combination of elements of Π with all coefficients nonpositive.

Let $\Pi = \{r_1, \ldots, r_m\}$ be a fixed t-base for Δ. A vector $x \in V$ is *t-positive* if it is possible to write x as a linear combination of $r_1, \ldots, r_m$ with all coefficients nonnegative. For example, every $r \in \Delta_t^+$ is t-positive. Similarly, $x \in V$ is *t-negative* if it is a nonpositive linear combination of $r_1, \ldots, r_m$. When there is no possibility of confusion we shall say *positive* rather than t-positive and *negative* rather than t-negative. Observe that if x is positive, then $(x, t) \geq 0$; and if x is negative, then $(x, t) \leq 0$. It will be shown (Proposition 4.1.8) that the t-base Π is unique, so the notion of positivity depends only on t and not on the choice of Π.

Proposition 4.1.4

If $r_i, r_j \in \Pi$, with $i \neq j$, and λ_i and λ_j are positive real numbers, then the vector $x = \lambda_i r_i - \lambda_j r_j$ is neither positive nor negative.

Proof

If x were positive, we could write

$$x = \lambda_i r_i - \lambda_j r_j = \Sigma_{k=1}^m u_k r_k$$

with all $\mu_k \geq 0$. If $\lambda_i \leq \mu_i$, then

$$0 = (\mu_i - \lambda_i) r_i + (\mu_j + \lambda_j) r_j + \Sigma\{\mu_k r_k : k \neq i, j\},$$

and so

$$\begin{aligned} 0 &= (t, (\mu_i - \lambda_i) r_i + (\mu_j + \lambda_j) r_j + \Sigma\{\mu_k r_k : k \neq i, j\}) \\ &\geq \lambda_j (t, r_j) > 0, \end{aligned}$$

a contradiction. If $\lambda_i > \mu_i$, then

$$(\lambda_i - \mu_i) r_i = (\mu_j + \lambda_j) r_j + \Sigma\{\mu_k r_k : k \neq i, j\}.$$

But then we may divide by $\lambda_i - \mu_i$ and express r_i as a nonnegative linear combination of the elements of $\Pi \setminus \{r_i\}$, contradicting the minimality of Π. Thus x is not positive. On the other hand, if x were negative then $-x$ would be positive, which is impossible by the above argument with i and j interchanged.

Proposition 4.1.5

Suppose that $r_i, r_j \in \Pi$, with $i \neq j$, and let S_i denote the reflection along r_i. Then $S_i r_j \in \Delta_t^+$, and $(r_i, r_j) \leq 0$.

Proof

Since $S_i r_j \in \Delta$, we know that $S_i r_j$ is either positive or negative. But

$$S_i r_j = r_j - 2(r_j, r_i) r_i,$$

with one coefficient positive. By Proposition 4.1.4 both coefficients must be nonnegative, so $(r_i, r_j) \le 0$ and $S_i r_j$ is positive.

Geometrically, $(r_i, r_j) \le 0$ means that the angle between the vectors r_i and r_j is obtuse, since (r_i, r_j) is the cosine of that angle.

Proposition 4.1.6

Suppose that $x_1, x_2, \dots, x_m \in V$ are all on the same side of a hyperplane; i.e., $(x_i, x) > 0$, $1 \le i \le m$, for some $x \in V$. If $(x_i, x_j) \le 0$ whenever $i \ne j$, then $\{x_1, \dots, x_m\}$ is a linearly independent set.

Proof

Suppose the contrary and relabel if necessary so that there is a dependence relation of the form

$$\Sigma_{i=1}^{k} \lambda_i x_i = \Sigma_{i=k+1}^{m} \mu_i x_i,$$

with all $\lambda_i \ge 0$, all $\mu_i \ge 0$, and some $\lambda_i > 0$. Then

$$\begin{aligned} 0 \le \|\Sigma_{i=1}^{k} \lambda_i x_i\|^2 &= (\Sigma_{i=1}^{k} \lambda_i x_i, \Sigma_{j=1}^{k} \lambda_j x_j) \\ &= (\Sigma_{i=1}^{k} \lambda_i x_i, \Sigma_{j=k+1}^{m} \mu_j x_j) \\ &= \Sigma_{i=1}^{k} \Sigma_{j=k+1}^{m} \lambda_i \mu_j (x_i, x_j) \le 0, \end{aligned}$$

so equality holds throughout. But then

$$0 = (\Sigma_{i=1}^{k} \lambda_i x_i, x) = \Sigma_{i=1}^{k} \lambda_i (x_i, x) > 0,$$

since some $\lambda_i > 0$. This is a contradiction and the proposition is proved.

Theorem 4.1.7

If Π is a t-base for Δ, then Π is a basis for V.

Proof

Since $\mathscr{G}$ is effective Δ spans V, by Proposition 4.1.2. Since every $r \in \Delta$ is a linear combination of roots in Π, V is spanned by Π. By Propositions 4.1.5 and 4.1.6, Π is linearly independent, so Π is a basis.

Proposition 4.1.8

There is only one t-base for Δ.

Proof

Suppose that Π_1 and Π_2 are t-bases. Since each root in Π_1 is a nonnegative linear combination of elements of Π_2, the change of basis matrix A from the basis Π_2 to the basis Π_1 has nonnegative entries. Likewise, the change of basis matrix $B = A^{-1}$ from Π_1 to Π_2 has nonnegative entries. Denote by $a_1, \ldots, a_n$ the rows of A and by $b_1, \ldots, b_n$ the columns of B. Since $AB = I$, we have $a_1^t \perp b_i$, $2 \leq i \leq n$, in $\mathscr{R}^n$. There can be at most one index j for which the jth entry in all of $b_2, \ldots, b_n$ is zero, for otherwise $b_2, \ldots, b_n$ would be linearly dependent, and hence B would be singular. It follows that a_1 has at most one nonzero entry. Similarly, each a_i has at most one nonzero entry. Since A is nonsingular, we conclude that A has exactly one positive entry in each row and in each column, and all other entries zero. Thus each root in Π_1 is a positive multiple of a root in Π_2. Since no positive multiple of a root r is a root except for r itself, A is a permutation matrix and $\Pi_1 = \Pi_2$.

When it is important to call attention to the vector t with respect to which positivity is defined, then the unique t-base for Δ will be denoted by Π_t. When such emphasis is unnecessary, however, we will usually write Π for Π_t, Δ^+ for Δ_t^+, and Δ^- for Δ_t^-.

In order to illustrate the concepts discussed thus far, let $\mathscr{G} = \mathscr{H}_2^4$, the dihedral group of order 8. The four reflections in $\mathscr{H}_2^4$ generate $\mathscr{H}_2^4$, and

$$\Delta = \{\pm(1, 0), \pm(0, 1), (\pm 1, \pm 1)\}.$$

Choosing $t = 2(\cos 3\pi/8, \sin 3\pi/8)$, we have

$$\Delta^+ = \{(1, 0), (1, 1), (0, 1), (-1, 1)\},$$

and

$$\Pi = \{(1, 0), (-1, 1)\}$$

(see Figure 4.2).

More generally, if $\mathscr{G} = \mathscr{H}_2^n$ and roots are taken to be unit vectors, then

$$\Delta = \{(\cos k\pi/n, \sin k\pi/n) : k = 0, 1, \ldots, 2n - 1\}.$$

Choosing $t = (\sin \pi/4n, \cos \pi/4n)$, we have

$$\Delta^+ = \{(\cos k\pi/n, \sin k\pi/n) : 0 \leq k \leq n - 1\},$$

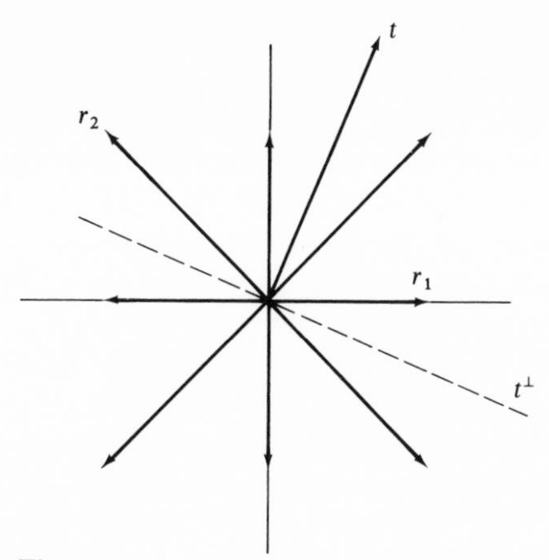

Figure 4.2

and

$$\Pi = \{(1, 0), (\cos(n - 1)\pi/n, \sin(n - 1)\pi/n)\}$$

(see Exercise 4.3).

Proposition 4.1.9

Suppose that S_i is the reflection along $r_i \in \Pi = \{r_1, \ldots, r_n\}$. If $r \in \Delta^+$ but $r \neq r_i$, then $S_i r \in \Delta^+$.

Proof

If $r \in \Pi$, then $S_i r \in \Delta^+$ by Proposition 4.1.5. If $r \notin \Pi$, then $r = \Sigma_j \lambda_j r_j$ and at least two of the coefficients λ_j are positive; so we may assume that $r_i \neq r_1$ and that $\lambda_1 > 0$. Thus

$$\begin{aligned} S_i r &= \Sigma_j \lambda_j S_i r_j \\ &= \lambda_1 r_1 + \Sigma_{j=2}^n \lambda_j r_j - 2(\Sigma_{j=1}^n \lambda_j (r_j, r_i)) r_i. \end{aligned}$$

Since $S_i r \in \Delta$, $S_i r$ is either positive or negative. Since it has at least one positive coefficient, λ_1, we conclude that all coefficients are nonnegative, and hence that $S_i r \in \Delta^+$.

The roots $r_1, \ldots, r_n$ in the base Π are sometimes called *fundamental roots*, or *simple roots*. The reflections $S_1, \ldots, S_n$ along the roots $r_1, \ldots, r_n$ will be called the *fundamental reflections* of $\mathcal{G}$. We shall temporarily denote by $\mathcal{G}_t$ the subgroup $\langle S_i : 1 \leq i \leq n \rangle$ of $\mathcal{G}$. It will be shown (Theorem 4.1.12) that $\mathcal{G}_t = \mathcal{G}$, i.e., that $\mathcal{G}$ is generated by its fundamental reflections.

Proposition 4.1.10

If $x \in V$, there is a transformation $T \in \mathscr{G}_t$ such that $(Tx, r_i) \geq 0$ for all $r_i \in \Pi$.

Proof

Set $x_0 = (1/2)\Sigma\{r : r \in \Delta^+\}$. Since $\mathscr{G}_t$ is a finite group, it is possible to choose $T \in \mathscr{G}_t$ for which (Tx, x_0) is maximal. If S_i is the reflection along r_i, then by Proposition 4.1.9 we have

$$\begin{aligned} S_i x_0 &= S_i((1/2)r_i + (1/2)\Sigma\{r \in \Delta^+ : r \neq r_i\}) \\ &= -(1/2)r_i + (1/2)\Sigma\{r \in \Delta^+ : r \neq r_i\} \\ &= 1/2\Sigma\{r : r \in \Delta^+\} - r_i = x_0 - r_i. \end{aligned}$$

Thus, by the maximality of (Tx, x_0),

$$\begin{aligned} (Tx, x_0) \geq (S_i Tx, x_0) &= (Tx, S_i x_0) = (Tx, x_0 - r_i) \\ &= (Tx, x_0) - (Tx, r_i); \end{aligned}$$

so $\qquad (Tx, r_i) > 0.$

Proposition 4.1.11

If $r \in \Delta^+$, then $Tr \in \Pi$ for some $T \in \mathscr{G}_t$.

If $r \in \Pi$, we may choose $T = 1$. If $r \notin \Pi$, then it follows from Propositions 4.1.5 and 4.1.6 and Theorem 4.1.7 that $(r, r_{i_1}) > 0$ for some root $r_{i_1} \in \Pi$; otherwise, $\Pi \cup \{r\}$ would be linearly independent. Set

$$a_1 = S_{i_1} r = r - 2(r, r_{i_1}) r_{i_1}.$$

Then $a_1 \in \Delta^+$ by Proposition 4.1.9, and

$$(a_1, t) = (r, t) - 2(r, r_{i_1})(r_{i_1}, t) < (r, t).$$

If $a_1 \in \Pi$, set $T = S_{i_1} \in \mathscr{G}_t$. If $a_1 \notin \Pi$, apply the above process to a_1, obtaining $r_{i_2} \in \Pi$ and

$$a_2 = S_{i_2} a_1 = S_{i_2} S_{i_1} r \in \Delta^+,$$

with $(a_2, t) < (a_1, t)$. If $a_2 \in \Pi$, set $T = S_{i_2} S_{i_1} \in \mathscr{G}_t$; if $a_2 \notin \Pi$, the process is continued. Since Δ^+ is finite, the process must terminate with some $a_k \in \Pi$. Since

$$a_k = S_{i_k} a_{k-1} = S_{i_k} \cdots S_{i_1} r,$$

the proposition is proved if we set $T = S_{i_k} \cdots S_{i_1} \in \mathscr{G}_t$.

Theorem 4.1.12

The fundamental reflections $S_1, \ldots, S_n$ generate $\mathscr{G}$; i.e., $\mathscr{G} = \mathscr{G}_t$.

Proof

Since $\mathcal{G} = \langle S_r : r \in \Delta \rangle$ and since $S_{-r} = S_r$, it will suffice to prove that if $r \in \Delta^+$, then $S_r \in \mathcal{G}_t$. Suppose then that $r \in \Delta^+$. By Proposition 4.1.11 there is a transformation $T \in \mathcal{G}_t$ such that $Tr \in \Pi$, say $Tr = r_i$. By Proposition 4.1.1 we have $S_r = T^{-1} S_i T \in \mathcal{G}_t$.

It may be worthwhile at this point to reflect momentarily on the progress we have made. For any Coxeter group $\mathcal{G}$ we have found a basis Π for V consisting of mutually obtuse roots whose reflections generate $\mathcal{G}$. Our procedure will be to obtain sufficient further geometrical information about the set Π in order to classify all possible t-bases for Coxeter groups, and thereby to classify the groups themselves.

4.2 FUNDAMENTAL REGIONS FOR COXETER GROUPS

Theorem 4.2.1

If $T \in \mathcal{G}$ and $T\Pi = \Pi$, then $T = 1$.

Proof

Suppose that $T \neq 1$. By Theorem 4.1.12 we may write T as $S_{i_1} S_{i_2} \cdots S_{i_k}$, a product of fundamental reflections. We may assume that T cannot be written as a product of fewer fundamental reflections, i.e., that k is minimal. Since $T \neq 1$, k is positive. Since $T\Pi = \Pi$, we have

$$Tr_{i_k} = S_{i_1} \cdots S_{i_k} r_{i_k} = -S_{i_1} \cdots S_{i_{k-1}} r_{i_k} \in \Pi,$$

so $S_{i_1} \cdots S_{i_{k-1}} r_{i_k} \in \Delta^-$. Set

$$\begin{aligned} a_0 &= S_{i_1} \cdots S_{i_{k-1}} r_{i_k}, \\ a_1 &= S_{i_1} a_0 = S_{i_2} \cdots S_{i_{k-1}} r_{i_k}, \\ a_2 &= S_{i_2} a_1 = S_{i_3} \cdots S_{i_{k-1}} r_{i_k}, \\ &\vdots \\ a_{k-1} &= S_{i_{k-1}} a_{k-2} = r_{i_k}, \end{aligned}$$

and observe that $a_0 \in \Delta^-$, $a_{k-1} \in \Delta^+$. Suppose that $a_0, a_1, \ldots, a_{j-1} \in \Delta^-$, but $a_j \in \Delta^+$; i.e., a_j is the first of the roots a_i to be positive. Since

$$a_j = S_{i_j} a_{j-1} \in \Delta^+,$$

and

$$S_{i_j} a_j = a_{j-1} \in \Delta^-,$$

it follows from Proposition 4.1.9 that $a_j = r_{i_j}$; so

$$r_{i_j} = S_{i_{j+1}} \cdots S_{i_{k-1}} r_{i_k}.$$

But then, by Proposition 4.1.1, we have

$$S_{i_j} = (S_{i_{j+1}} \cdots S_{i_{k-1}})S_{i_k}(S_{i_{j+1}} \cdots S_{i_{k-1}})^{-1}.$$

Thus

$$S_{i_j}(S_{i_{j+1}} \cdots S_{i_{k-1}}) = (S_{i_{j+1}} \cdots S_{i_{k-1}})S_{i_k},$$

and so

$$\begin{aligned} T = S_{i_1} \cdots S_{i_k} &= (S_{i_1} \cdots S_{i_{j-1}})(S_{i_j} \cdots S_{i_{k-1}})S_{i_k} \\ &= (S_{i_1} \cdots S_{i_{j-1}})(S_{i_{j+1}} \cdots S_{i_k})S_{i_k} \\ &= S_{i_1} \cdots S_{i_{j-1}} S_{i_{j+1}} \cdots S_{i_{k-1}}, \end{aligned}$$

representing T as a product of $k - 2$ fundamental reflections and contradicting the fact that k was minimal.

Proposition 4.2.2
If $T \in \mathcal{G}$, then $T(\Delta_t^+) = \Delta_{T(t)}^+$; consequently, $T(\Pi_t) = \Pi_{T(t)}$.

Proof
Since every root in $T(\Delta_t^+)$ is a nonnegative linear combination of roots in $T(\Pi_t)$, the second statement follows from the first by Proposition 4.1.8. As for the first statement,

$$\begin{aligned} T(\Delta_t^+) &= T\{r \in \Delta : (t, r) > 0\} \\ &= \{Tr \in \Delta : (t, r) = (Tt, Tr) > 0\} \\ &= \{s \in \Delta : (Tt, s) > 0\} = \Delta_{T(t)}^+. \end{aligned}$$

Proposition 4.2.3
If $T \in \mathcal{G}$ and $T(\Delta^+) = \Delta^+$, then $T = 1$.

Proof
By Proposition 4.2.2 we have

$$\Delta_t^+ = T(\Delta_t^+) = \Delta_{T(t)}^+,$$

so $\Pi_t = \Pi_{T(t)}$ by Proposition 4.1.8. But then $T\Pi_t = \Pi_t$ by Proposition 4.2.2, so $T = 1$ by Theorem 4.2.1.

Let us denote by $\Pi^* = \{s_1, \ldots, s_n\}$ the dual basis of Π in V, so that $(r_i, s_j) = \delta_{ij}$ for all i and j. Set

$$F_t = \{x \in V : x = \Sigma_{i=1}^n \lambda_i s_i, \lambda_i \in \mathcal{R}, \text{all } \lambda_i > 0\}.$$

As usual, we will normally suppress the dependence of F_t on the vector t and simply write F when there is no danger of confusion.

For any $x \in V$, write $x = \Sigma_{i=1}^n \lambda_i s_i$. Then

$$(x, r_j) = \Sigma_i \lambda_i (s_i, r_j) = \lambda_j$$

for all j, so $x = \Sigma_i (x, r_i) s_i$. Thus

$$\begin{aligned} F_t = F &= \{x \in V : (x, r_i) > 0, \text{ all } r_i \in \Pi\} \\ &= \textstyle\bigcap_{i=1}^n \{x \in V : (x, r_i) > 0\}. \end{aligned}$$

In other words, F is the intersection of open half-spaces determined by the hyperplanes $\mathscr{P}_i = r_i^\perp$, $r_i \in \Pi$. It follows that F is open and convex. Also, F^- is the intersection of the closed half-spaces $\{x \in V : (x, r_i) \geq 0\}$, and the boundary of F is the union of the intersections with F^- of the hyperplanes $\mathscr{P}_i$. The subsets $F^- \cap \mathscr{P}_i$ of the boundary are called the *walls* of F, and we shall speak of the fundamental reflection S_i through $\mathscr{P}_i$ as being a reflection through the ith wall of F.

Theorem 4.2.4

The set $F = F_t$ is a fundamental region for the Coxeter group $\mathscr{G}$.

Proof

It was observed above that F is open. Suppose that $T \in \mathscr{G}$ and $x \in F \cap TF$. Setting $R = T^{-1}$, we have $Rx = T^{-1} x \in F$, since $x \in TF$. Since $x \in F$, $(x, r_i) > 0$ for all i; and so $(x, r) > 0$ for all $r \in \Delta_t^+$. It follows immediately that $\Delta_x^+ = \Delta_t^+$, and hence that $\Pi_x = \Pi_t$, by Proposition 4.1.8. The same reasoning shows that $\Pi_{Rx} = \Pi_t$. Using Proposition 4.2.2, we have

$$\Pi_t = \Pi_{Rx} = R\Pi_x = R\Pi_t.$$

Thus $R = T = 1$ by Theorem 4.2.1. Finally, if $y \in V$, then by Proposition 4.1.10 there is a transformation $T \in \mathscr{G}$ such that $(Ty, r_i) \geq 0$ for all $r_i \in \Pi$, and so $Ty \in F^-$. Thus

$$y \in T^{-1}(F^-) = (T^{-1}F)^-;$$

so

$$V = \bigcup \{(RF)^- : R \in \mathscr{G}\}$$

and the theorem is proved.

Let us give one further characterization of F^-. The *convex hull* of the vectors $s_1, s_2, \ldots, s_n \in \Pi^*$ is by definition the smallest convex subset of

V containing Π^*. Thus the convex hull of Π^* is

$$\{x \in V : x = \Sigma_{i=1}^n \lambda_i s_i, \lambda_i \geq 0, \Sigma_i \lambda_i = 1\}$$

(see Exercise 4.13). If we denote the convex hull of Π^* by $\mathrm{co}(\Pi^*)$, then

$$F^- = \bigcup\{\lambda \,\mathrm{co}(\Pi^*) : 0 \leq \lambda \in \mathscr{R}\};$$

i.e., F^- is the (scalar) product of the closed half-line $[0, \infty)$ with the subset $\mathrm{co}(\Pi^*)$ of V. The set $\mathrm{co}(\Pi^*)$ is sometimes called the *simplex* spanned by $\{s_1, \ldots, s_n\}$, and the product $[0, \infty) \cdot \mathrm{co}(\Pi^*)$ is the corresponding *simplicial cone*.

We may summarize as follows: A Coxeter group $\mathscr{G}$ has a fundamental region F whose closure F^- is a simplicial cone, and $\mathscr{G}$ is generated by the reflections through the walls of that fundamental region.

Theorem 4.2.5

Every reflection in $\mathscr{G}$ is conjugate in $\mathscr{G}$ to a fundamental reflection; consequently, every root of $\mathscr{G}$ is in the root system Δ.

Proof

Suppose that $S_r \in \mathscr{G}$ is a reflection, with root r, and set $\mathscr{P} = r^\perp$. If F is the fundamental region discussed above, then each TF, $T \in \mathscr{G}$, is a fundamental region for $\mathscr{G}$. If $\mathscr{P} \cap TF \neq \varnothing$ for some $T \in \mathscr{G}$, choose $x \in \mathscr{P} \cap TF$. Since TF is open, the ball B of radius ε centered at x lies entirely within TF for some sufficiently small $\varepsilon > 0$. Since $S_r x = x$ and S_r preserves distances, we have $S_r B = B$. But also $B \nsubseteq \mathscr{P}$ (see Exercise 4.14), so we may choose $y \in B \setminus \mathscr{P}$. Then $S_r y \in B \subseteq TF$, but $S_r y \neq y$, which is in conflict with the fact that TF is a fundamental region. Thus

$$\begin{aligned} \mathscr{P} &\subseteq V \setminus \bigcup\{TF : T \in \mathscr{G}\} \\ &= \bigcup\{T\mathscr{P}_i : T \in \mathscr{G}, 1 \leq i \leq n\}; \end{aligned}$$

$$\mathscr{P} = \bigcup\{\mathscr{P} \cap T\mathscr{P}_i : T \in \mathscr{G}, 1 \leq i \leq n\}.$$

As a result, $\mathscr{P} = \mathscr{P} \cap T\mathscr{P}_i$, or $\mathscr{P} \subseteq T\mathscr{P}_i$, for some $T \in \mathscr{G}$ and some i, by Proposition 3.1.1. Since both $\mathscr{P}$ and $T\mathscr{P}_i$ are hyperplanes, we may conclude that $\mathscr{P} = T\mathscr{P}_i$. But then either $r = Tr_i$ or $r = -Tr_i = TS_i r_i$. In either case $r \in \Delta$ and $S_r = TS_i T^{-1}$.

Let us illustrate the above results. Suppose first that

$$\mathscr{G} = \mathscr{H}_2^3 \leq \mathscr{O}(\mathscr{R}^2).$$

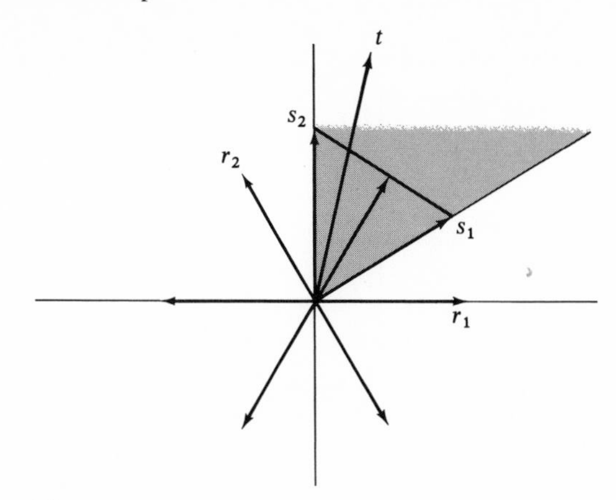

Figure 4.3

Referring to Figure 4.3, we have chosen $t = (1/2, 2)$ so that

$$\Delta^+ = \{(1, 0), (1/2, \sqrt{3}/2), (-1/2, \sqrt{3}/2)\},$$

and

$$\Pi = \{r_1, r_2\} = \{(1, 0), (-1/2, \sqrt{3}/2)\}.$$

Then $s_1 = (1, \sqrt{3}/3)$ and $s_2 = (0, 2\sqrt{3}/3)$. The convex hull of $\Pi^* = \{s_1, s_2\}$ is just the line segment $[s_1 s_2]$, and the shaded region is the fundamental region F.

Suppose next that $\mathscr{G} = \mathscr{W}^* \leq \mathscr{O}(\mathscr{R}^3)$ is the group of symmetries of the cube. Suppose that the cube is situated with its center at the origin and its vertices at the eight points $(\pm 1, \pm 1, \pm 1)$. Then the roots of $\mathscr{G}$ are $\{\pm r_1, \ldots, \pm r_9\}$, where

$$r_1 = e_1, \qquad r_2 = e_2 - e_1, \qquad r_3 = e_3 - e_2,$$

$$r_4 = e_2, \qquad r_5 = e_3 - e_1, \qquad r_6 = e_1 + e_2,$$

$$r_7 = e_3, \qquad r_8 = e_2 + e_3, \qquad r_9 = e_1 + e_3.$$

If we choose $t = (1, 2, 3)$, then

$$\Delta^+ = \{r_1, \ldots, r_9\} \qquad \text{and} \qquad \Pi = \{r_1, r_2, r_3\}.$$

The dual basis is $\Pi^* = \{s_1, s_2, s_3\}$, with

$$s_1 = e_1 + e_2 + e_3, \qquad s_2 = e_2 + e_3, \qquad s_3 = e_3.$$

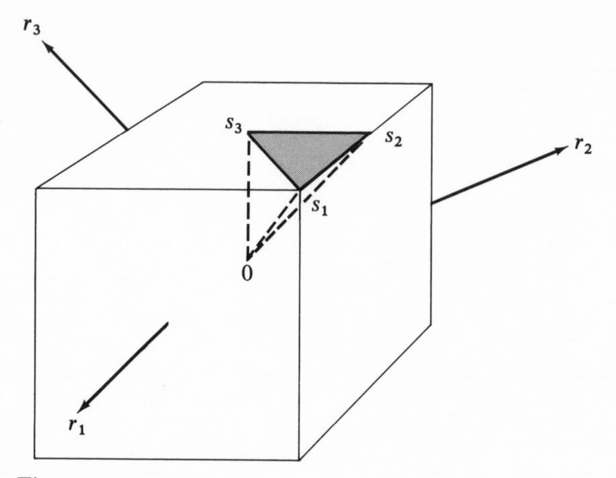

Figure 4.4

The cube is shown in Figure 4.4 with the vectors r_1, r_2, and r_3 displaced so that they emanate from the surface of the cube rather than from the origin. The shaded region is the intersection of the fundamental region F with the surface of the cube.

For an example of a different sort let $\mathscr{G} = \mathscr{I}^*$, the symmetry group of an icosahedron. By Exercise 2.23 we may take the vectors

$$a = (1, 0, 2\alpha), \qquad b = (0, 2\alpha, 1), \qquad c = (2\alpha, 1, 0)$$

as the vertices of one face of the icosahedron, since they are at distance 2 from one another. An application of the Fricke–Klein construction, as in Chapter 3, shows that the interior of the triangle with vertices

$$a_1 = \frac{a + b + c}{3}, \qquad a_2 = \frac{a + b}{2}, \qquad a_3 = a$$

is a fundamental region for $\mathscr{I}^*$ in the surface of the icosahedron (see Figure 4.5). If the vector t is chosen within that triangle, then the interior of the triangle is the intersection of F_t with the surface of the icosahedron (see Exercise 4.10). If $\{r_1, r_2, r_3\}$ is the set of fundamental roots of $\mathscr{I}^*$ and $\{s_1, s_2, s_3\}$ is the dual basis, then the edges of the simplicial cone F^- are spanned by $\{s_1, s_2, s_3\}$ as well as by $\{a_1, a_2, a_3\}$. Thus we may assume that $s_i = \gamma_i a_i$, $i = 1, 2, 3$, where each $\gamma_i > 0$.

It is now possible to determine the fundamental roots r_i, if we require that $\|r_i\| = 1$. For example, suppose that $r_1 = (\lambda, \mu, \nu)$. Then $r_1 \perp a_2$

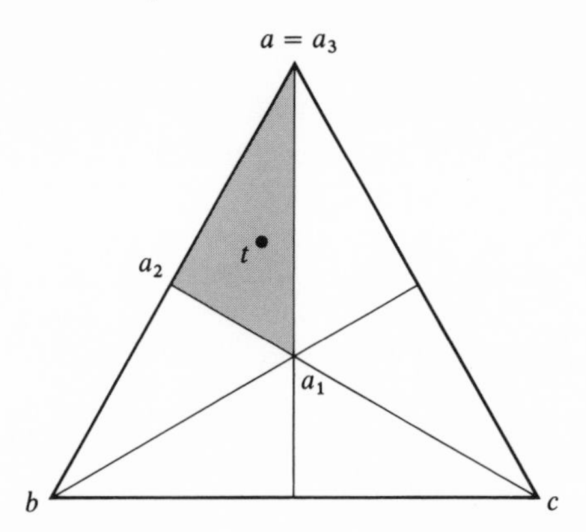

Figure 4.5

and $r_1 \perp a_3$, yielding the equations

$$\begin{aligned} \lambda + 2\alpha\mu + (2\alpha + 1)\nu &= 0 \\ \lambda \qquad\qquad + 2\alpha\nu &= 0. \end{aligned}$$

Solving, we have

$$r_1 = \nu(-2\alpha, 1 - 2\alpha, 1);$$

$$1 = \|r_1\|^2 = \nu^2(8\alpha^2 - 4\alpha + 2) = 4\nu^2,$$

or $\nu = \pm 1/2$ (see Exercise 2.22). Since $(r_1, a_1) = \gamma_1^{-1} > 0$ we see that $\nu = -1/2$ and $\gamma_1 = 6\beta$. Note that

$$r_1 = (-1/2)(-2\alpha, 1 - 2\alpha, 1) = \beta(2\alpha + 1, 1, -2\alpha).$$

Similar computations determine r_2 and r_3, and we have

$$\begin{aligned} r_1 &= \beta(2\alpha + 1, 1, -2\alpha), \\ r_2 &= \beta(-2\alpha - 1, 1, 2\alpha), \\ r_3 &= \beta(2\alpha, -2\alpha - 1, 1), \\ s_1 &= 6\beta a_1 = (2\alpha, 2\alpha, 2\alpha), \\ s_2 &= 2a_2 = (1, 2\alpha, 2\alpha + 1), \\ s_3 &= a_3 = (1, 0, 2\alpha). \end{aligned}$$

The final theorem of this chapter is not essential to the development that follows. It is presented to give further geometrical insight into the fundamental regions constructed above.

Theorem 4.2.6

Suppose that $\{r_1, \ldots, r_n\}$ is a basis for V with $(r_i, r_j) \leq 0$ if $i \neq j$, and let $\{s_1, \ldots, s_n\}$ be the dual basis. Then $(s_i, s_j) \geq 0$ for all i, j.

Proof (R. Koch and T. Matthes)

Let A be the matrix whose ijth entry is (r_i, r_j) and B the matrix whose ijth entry is (s_i, s_j). Then $B = A^{-1}$ (see Exercise 4.15), and we must show that B has nonnegative entries. Since $\{r_1, \ldots, r_n\}$ is a basis, the matrix A is positive definite (for a proof see the proof of Theorem 5.1.3). Since each diagonal entry of A is $\|r_i\|^2 = 1$, the trace of A is n. Thus every eigenvalue of A is positive and strictly less than n. It follows that if λ is an eigenvalue of $I - (1/n)A$, then $0 < \lambda < 1$. Setting $C = I - (1/n)A$, we have $A = n(I - C)$; so

$$A^{-1} = (1/n)(I - C)^{-1} = (1/n)(I + C + C^2 + \cdots)$$

(the series converges, entry by entry, since the eigenvalues of C are positive and less than 1). But all entries of C are nonnegative, since $(r_i, r_j) \leq 0$ for $i \neq j$; so all entries of B are nonnegative, and the theorem is proved.

The half-line $l_i = \{\lambda_{s_i} : \lambda \in \mathscr{R}, \lambda \geq 0\}$ is the intersection of all walls except the ith of the simplicial cone F^- (see Exercise 4.16). By analogy with the three-dimensional case we may think of l_i as being the *edge* of F opposite the ith face. With this interpretation Theorem 4.2.6 says that the edges of the fundamental region F are all at acute angles with one another.

Exercises

4.1 If $0 \neq r \in V$ and S_r is defined by the formula

$$S_r x = x - \frac{2(x, r)r}{(r, r)}$$

for all $x \in V$, show that $S_r \in \mathcal{O}(V)$.

4.2 In Proposition 4.1.3 show that $\mathscr{G}$ is finite if Δ is finite, even if $\mathscr{G}$ is *not* effective.

4.3 (a) Verify the statements made concerning root systems and t-bases for the groups $\mathscr{H}_2^n$ on pages 40 and 41.
(b) Find the dual basis Π^* for each $\mathscr{H}_2^n$.

4.4 Suppose that Π is a t-base for Δ.
(a) If $r \in \Delta^+$, show that $r \in \Pi$ if and only if r is not a (strictly) positive linear combination of two or more positive roots.
(b) Use (a) to give an alternate proof of the uniqueness of Π.

4.5 Suppose that Π_t and Π_s are a t-base and an s-base, respectively, for Δ. Show that $T\Pi_s = \Pi_t$ for some $T \in \mathscr{G}$.

4.6 Prove that $\mathscr{H}_2^n$ $(n \geq 1)$, $\mathscr{W}^*$, $\mathscr{W}]\mathscr{T}$, and $\mathscr{I}^*$ are all Coxeter groups.

4.7 Verify all the statements made about $\mathscr{W}^*$ in the example on pages 47 and 48.

4.8 Set $r_1 = e_1 - e_2$, $r_2 = e_2 - e_3$, and $r_3 = -e_1 - e_2$ in $\mathscr{R}^3$. If $t = e_1 - 2e_2 - 3e_3$, then $\{r_1, r_2, r_3\}$ is a t-base for $\mathscr{W}]\mathscr{T}$.

(a) Write the matrices representing S_1, S_2, and S_3 with respect to the basis $\{r_1, r_2, r_3\}$ (*Note:* The matrices will *not* be orthogonal matrices since the basis is not orthonormal).

(b) Form products and write matrices representing all 24 transformations in $\mathscr{W}]\mathscr{T}$.

(c) Find the root system of $\mathscr{W}]\mathscr{T}$.

(d) Find the axes of rotation of the rotations of order 3 [i.e., find eigenvectors with eigenvalue 1 (see Euler's theorem)]; and hence find vertices for a tetrahedron left invariant by $\mathscr{W}]\mathscr{T}$.

(e) Find the dual basis Π^* and the fundamental region F for $\mathscr{W}]\mathscr{T}$.

4.9 Give an example of an infinite group generated by two reflections in $\mathscr{O}(\mathscr{R}^2)$.

4.10 Show that the fundamental region F_t may be obtained by the Fricke–Klein construction of Chapter 3, choosing $x_0 = t$.

4.11 Show that $F_t = \{u \in V : \Pi_u = \Pi_t\}$.

4.12 Provide the details of the proofs that

$$F^- = \bigcap_{i=1}^n \{x : (x, r_i) \geq 0\}$$

and that the boundary of F is $\bigcup_{i=1}^n (F^- \cap \mathscr{P}_i)$.

4.13 (a) If $X \subseteq V$, show that there is a smallest convex set containing X [i.e., a convex hull co(X)] by showing that the intersection of all convex sets containing X is convex.

(b) Show that

$$\text{co}(X) = \{\Sigma_{i=1}^m \lambda_i x_i : m \geq 1, \lambda_i > 0, x_i \in X, \Sigma_i \lambda_i = 1\}.$$

4.14 If $0 < \varepsilon \in \mathscr{R}$, let B_ε denote the ball in V of radius ε, centered at the origin.

(a) For any $x \in V$, show that V has a basis $\{x_1, \ldots, x_n\} \subseteq B_\varepsilon$ such that $-x \neq \Sigma_i \lambda_i x_i$, with $\Sigma_i \lambda_i = 1$.

(b) Use (a) to show that $\{x_1 + x, \ldots, x_n + x\}$ is a basis for V contained in the ball B of radius ε centered at x.

(c) Conclude that no proper subspace of V contains a nonempty set that is open in V.

(d) Readers familiar with the Baire Category Theorem can use (c) to prove a strengthened version of Proposition 3.1.1.

4.15 Suppose that $\{x_1, \ldots, x_n\}$ is a basis for V and $\{y_1, \ldots, y_n\}$ is the dual basis. Let A be the matrix with ijth entry (x_i, x_j) and let B be the matrix with ijth entry (y_i, y_j).
(a) Show that $x_i = \Sigma_j(x_i, x_j)y_j$ and $y_i = \Sigma_j(y_i, y_j)x_j$ for each i.
(b) Conclude that $B = A^{-1}$.

4.16 Prove that the "positive" half-line spanned by $s_i \in \Pi^*$ is the intersection of all walls except the ith of the fundamental region F.

4.17 If $\dim V = n$ and $\{x_1, x_2, \ldots, x_{n+2}\} \subseteq V$, show that $(x_i, x_j) \geq 0$ for some $i \neq j$.

4.18 If $x, y \in V$, with $(x, r_i) > 0$ and $(y, r_i) > 0$ for all $r_i \in \Pi$, show that $(x, y) \geq 0$. Use this fact to give an alternate proof of Theorem 4.2.6.

chapter 5

CLASSIFICATION OF COXETER GROUPS

5.1 COXETER GRAPHS

We continue to assume that $\mathscr{G} \leq \mathcal{O}(V)$ is a Coxeter group with root system Δ, and that $\Pi = \{r_1, \ldots, r_n\}$ is a t-base for some $t \in V$. As before, the fundamental reflection along r_i will be denoted by S_i.

Proposition 5.1.1

If $r_i, r_j \in \Pi$, then there is an integer $p_{ij} \geq 1$ such that

$$(r_i, r_j)/\|r_i\| \, \|r_j\| = -\cos(\pi/p_{ij}).$$

In fact, p_{ij} is the order of $S_i S_j$ as a group element.

Proof

If $i = j$, we may take $p_{ij} = 1$. Assume then that $i \neq j$, and denote by W the two-dimensional subspace of V spanned by r_i and r_j. Let $\mathscr{H}$ be the subgroup of $\mathscr{G}$ spanned by S_i and S_j. Since $S_i|W^{\perp} = S_j|W^{\perp} = 1$, $\mathscr{H}$ has the form of a direct product $\mathscr{H}_2^m \times 1$, where $\mathscr{H}_2^m$ is a dihedral group in $\mathcal{O}(W)$. If we write $t = t_1 + t_2$, with $t_1 \in W$ and $t_2 \in W^{\perp}$, let us show that $\{r_i, r_j\}$ is a t_1-base for $\mathscr{H}_2^m$ in W. If it is not, then we may choose a root r of $\mathscr{H}_2^m$ in W such that $\{r, r_j\}$ is a t_1'-base for $\mathscr{H}_2^m$ for some $t_1' \in W$, where r, r_i, and r_j are all t_1'-positive (see Figure 5.1). Considered as a vector in V, r is a root of $\mathscr{H}$, and hence of $\mathscr{G}$. If we express r as a linear combination of r_i and r_j, then clearly it is of the form $\lambda_i r_i - \lambda_j r_j$ with $\lambda_i > 0$ and $\lambda_j > 0$. This contradicts Proposition 4.1.4 since r, being a root of $\mathscr{G}$, must be either t-positive or t-negative.

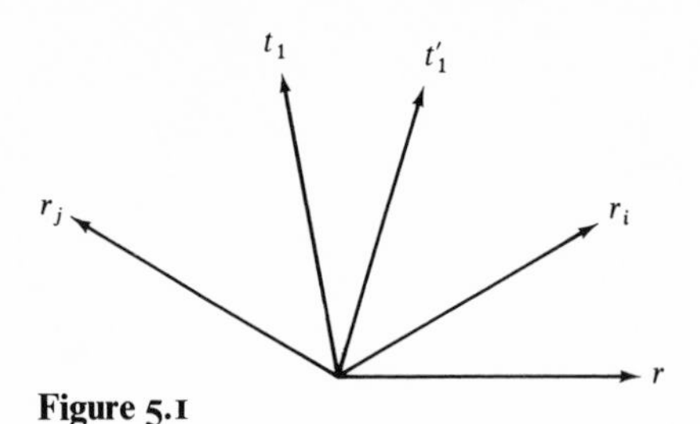

Figure 5.1

We may ignore the factor $1/\|r_1\|\ \|r_2\|$, since we are still assuming that all roots are unit vectors. The acute angle φ between the basis vectors dual to the fundamental roots r_i and r_j in W must be $2\pi/2m$, since the interior of the simplicial cone they span is a fundamental region for $\mathscr{H}_2^m$. If θ is the least angle between r_i and r_j, then $\theta = \pi - \varphi$. Set $p_{ij} = m$. Then

$$(r_i, r_j) = \cos\theta = \cos(\pi - \varphi)$$
$$= -\cos\varphi = -\cos(\pi/p_{ij}).$$

The second statement of the proposition is a consequence of Exercise 2.5.

A *marked graph* is a finite set of points (called *nodes*) such that any two distinct points may or may not be joined by a line (called a *branch*), with the following property: If there is a branch joining the ith and jth nodes, then that branch is marked, or labeled, with a real number $p_{ij} > 2$. If G is a marked graph for which every mark p_{ij} is an integer, then G is called a *Coxeter graph.*

If G is a marked graph, then as a matter of convenience we shall usually suppress the label on any branch for which $p_{ij} = 3$.

Readers familiar with the theory of Lie algebras will observe similarities between the notions of Coxeter graphs and Dynkin diagrams. We remark in passing that a Dynkin diagram is a Coxeter graph with the further restriction that $p_{ij} = 3$, 4, or 6, in which branches marked 4 are replaced by double branches and branches marked 6 are replaced by triple branches.

If G is a marked graph with m nodes, we may associate with G a quadratic form $Q = Q_G$ on $\mathscr{R}^m$ as follows:

$$Q(\lambda_1, \ldots, \lambda_m) = \Sigma_{i,j}\ \alpha_{ij}\lambda_i\lambda_j,$$

where $A = (\alpha_{ij})$ is the symmetric matrix with $\alpha_{ii} = 1$, $\alpha_{ij} = -\cos(\pi/p_{ij})$ if there is a branch joining the ith and jth nodes of G, and $\alpha_{ij} = 0$ otherwise. Note that we may write $\alpha_{ij} = -\cos(\pi/p_{ij})$ for all i and j if we agree that $p_{ii} = 1$ for all i, and that $p_{ij} = 2$ whenever there is no branch joining the

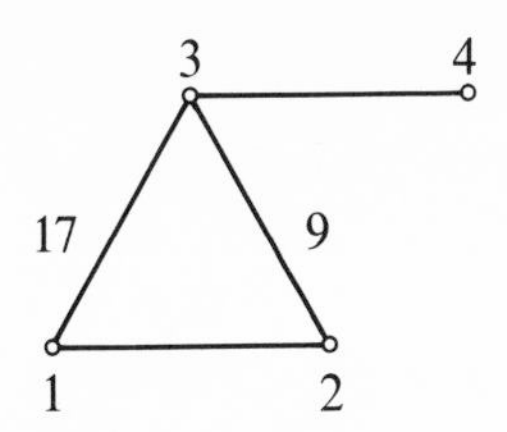

Figure 5.2

ith and jth nodes. If we set $x = (\lambda_1, \ldots, \lambda_m) \in \mathscr{R}^m$, observe that $Q(x) = (Ax, x)$.

A marked graph will be called *positive definite* if and only if its associated quadratic form is positive definite.

The graph G in Figure 5.2 is an example of a Coxeter graph, with the nodes labeled from 1 to 4. The matrix defining the associated quadratic form on $\mathscr{R}^4$ is

$$A = \begin{bmatrix} 1 & -1/2 & -\cos(\pi/17) & 0 \\ -1/2 & 1 & -\cos(\pi/9) & 0 \\ -\cos(\pi/17) & -\cos(\pi/9) & 1 & -1/2 \\ 0 & 0 & -1/2 & 1 \end{bmatrix}.$$

If $\{x_1, x_2, \ldots, x_m\}$ is any finite set of mutually obtuse vectors in V, we may define a marked graph G as follows: Let G have m nodes, and if $i \neq j$ the ith and jth nodes are joined by a branch if and only if $(x_i, x_j) \neq 0$. In that case we may write $(x_i, x_j)/\|x_i\| \, \|x_j\| = -\cos(\pi/p_{ij})$ for exactly one real number $p_{ij} > 2$, and the branch is labeled p_{ij}.

If $\mathscr{G}$ is a Coxeter group, then the marked graph G corresponding to the t-base $\Pi = \{r_1, \ldots, r_n\}$ is a Coxeter graph, by Proposition 5.1.1, and G is called the Coxeter graph of $\mathscr{G}$.

Theorem 5.1.2

If $\mathscr{G}_1, \mathscr{G}_2 \leq \mathscr{O}(V)$ are Coxeter groups having the same Coxeter graph, then $\mathscr{G}_1$ and $\mathscr{G}_2$ are geometrically the same, in the sense that $T\mathscr{G}_1 T^{-1} = \mathscr{G}_2$ for some transformation $T \in \mathscr{O}(V)$.

Proof

Let Π_1 and Π_2 be bases of unit vectors for the root systems of $\mathscr{G}_1$ and $\mathscr{G}_2$, respectively. To say that the two groups have the same Coxeter graph simply means that the elements of Π_1 and Π_2 may be written as

$$\Pi_1 = \{r_1, \ldots, r_n\} \qquad \text{and} \qquad \Pi_2 = \{r_1', \ldots, r_n'\}$$

with $(r_i, r_j) = (r_i', r_j')$ for all i and j. Define $T: V \to V$ by setting $Tr_i = r_i'$, $1 \leq i \leq n$, and extending by linearity. Then $T \in \mathcal{O}(V)$ (see Exercise 5.2). If $S_i \in \mathcal{G}_1$ is the reflection along r_i and $S_i' \in \mathcal{G}_2$ is the reflection along r_i', then $S_i' = TS_iT^{-1}$ for each i by Proposition 4.1.1 [with $\mathcal{G} = \mathcal{O}(V)$]. Since $\mathcal{G}_1$ is generated by the reflections S_i and $\mathcal{G}_2$ by the reflections S_i', the conclusion follows easily.

Theorem 5.1.3
The Coxeter graph of a Coxeter group is positive definite.

Proof
If roots are taken to be unit vectors, then the matrix A defining the associated quadratic form has ijth entry (r_i, r_j). If $0 \neq x = (\lambda_1, \ldots, \lambda_n)$ in $\mathcal{R}^n$, then $\Sigma_i \lambda_i r_i \neq 0$ in V since Π is linearly independent. Thus

$$\begin{aligned} Q(x) &= \Sigma_{i,j}(r_i, r_j)\lambda_i\lambda_j \\ &= (\Sigma_i \lambda_i r_i, \Sigma_j \lambda_j r_j) \\ &= \|\Sigma_i \lambda_i r_i\|^2 > 0; \end{aligned}$$

so Q is positive definite.

Suppose that Π is the t-base of $\mathcal{G}$ and that $\Pi = \Pi_1 \cup \Pi_2$, with Π_1 and Π_2 nonempty and $\Pi_1 \perp \Pi_2$. Let V_i be the subspace of V spanned by Π_i, so that $V_1 \perp V_2$ and $V = V_1 \oplus V_2$. If $r_i \in \Pi_1$ and $r_j \in \Pi_2$, then $S_i r_j = r_j$ since $(r_i, r_j) = 0$. Thus the restriction $S_i | V_2$ is the identity transformation on V_2; and, in particular, $S_i V_2 = V_2$. Since $V_1 = V_2^{\perp}$, we also have $S_i V_1 = V_1$ for all $r_i \in \Pi_1$. Similarly, $S_j | V_1$ is the identity on V_1 and $S_j V_2 = V_2$ for all $r_j \in \Pi_2$. It follows from Theorem 4.1.12 that $TV_1 = V_1$ and $TV_2 = V_2$ for all $T \in \mathcal{G}$. Set $\mathcal{G}_1 = \{T|V_1 : T \in \mathcal{G}\} \leq \mathcal{O}(V_1)$ and set $\mathcal{G}_2 = \{T|V_2 : T \in \mathcal{G}\} \leq \mathcal{O}(V_2)$. Then $\mathcal{G}_1$ and $\mathcal{G}_2$ are Coxeter subgroups of $\mathcal{O}(V_1)$ and $\mathcal{O}(V_2)$, with $\mathcal{G}_1$ generated by the reflections $S_i|V_1$ along the roots $r_i \in \Pi_1$, and $\mathcal{G}_2$ generated by the reflections $S_j|V_2$, $r_j \in \Pi_2$. Furthermore, each T can be expressed as $T|V_1 \oplus T|V_2$ acting on $V_1 \oplus V_2$; so $\mathcal{G}$ is isomorphic with $\mathcal{G}_1 \times \mathcal{G}_2$, and the study of $\mathcal{G}$ has been reduced to the study of the smaller Coxeter groups $\mathcal{G}_1$ and $\mathcal{G}_2$.

If Π is *not* the union of two nonempty orthogonal subsets, we shall say that $\mathcal{G}$ is *irreducible*. Otherwise $\mathcal{G}$ is called *reducible*. The crux of the discussion above is that we shall lose no generality if we restrict attention to irreducible Coxeter groups.

Two distinct nodes a and b in a marked graph G are said to be *connected in* G if and only if there are nodes $a_1, \ldots, a_k$ in G such that $a_1 = a$, a_1 and a_2 are joined by a branch, a_2 and a_3 are joined by a branch, ...,

a_{k-1} and a_k are joined by a branch, and $a_k = b$. If every two distinct nodes of G are connected in G, then G is said to be *connected*. For example, the graph in Figure 5.2 is connected.

Proposition 5.1.4 is an immediate consequence of the foregoing definitions.

Proposition 5.1.4

The Coxeter graph of a Coxeter group $\mathscr{G}$ is connected if and only if $\mathscr{G}$ is irreducible.

In order to arrive eventually at a classification of all Coxeter subgroups of $\mathcal{O}(V)$, we shall classify all positive definite Coxeter graphs. As we observed above, it will suffice to consider only irreducible Coxeter groups, so we need only classify the connected positive definite Coxeter graphs.

Consider first the Coxeter graphs shown in Figure 5.3. The subscripts on the names of the graphs indicate the numbers of nodes. The graph H_2^n is clearly the Coxeter graph of the dihedral group $\mathscr{H}_2^n$, so it is positive definite by Theorem 5.1.3. Note that $\mathscr{H}_2^2$ is a reducible group and that its Coxeter graph is simply two nodes with no branch joining them. To avoid repetitions we have not included the graphs of $\mathscr{H}_2^3$, $\mathscr{H}_2^4$, and $\mathscr{H}_2^6$ among the graphs H_2^n, since their graphs are listed as A_2, B_2, and G_2. It will become apparent in Section 5.2 why the graph of $\mathscr{H}_2^6$ has been singled out as G_2.

If $1 \leq k \leq n$, then the kth *principal minor* of an $n \times n$ matrix A is the determinant of the $k \times k$ matrix obtained by deleting the last $n - k$

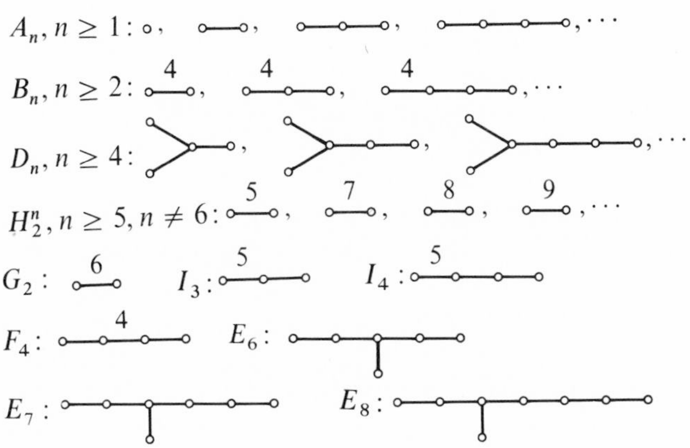

$A_n, n \geq 1$: ...

$B_n, n \geq 2$: 4, 4, 4, ...

$D_n, n \geq 4$: ...

$H_2^n, n \geq 5, n \neq 6$: 5, 7, 8, 9, ...

G_2: 6 $\quad I_3$: 5 $\quad I_4$: 5

F_4: 4 $\quad E_6$:

E_7: $\quad E_8$:

Figure 5.3

rows and columns of A. For example, the first principal minor of A is the first diagonal entry, and the nth principal minor is $\det A$. For a real symmetric matrix (or its associated quadratic form) to be positive definite it is necessary and sufficient that all its principal minors be positive (for a proof see [14], p. 152, or [17], p. 167). We shall apply this criterion in order to show that the remaining graphs in Figure 5.3 are positive definite.

For each Coxeter graph G the matrix of the associated quadratic form can be written down directly from the graph. It will be convenient to denote the matrix by the same name as the graph in each case.

The matrix of A_n is

$$\left[\begin{array}{ccccccccc} 1 & -1/2 & 0 & 0 & \cdots & & & & \\ -1/2 & 1 & -1/2 & 0 & \cdots & & 0 & & \\ 0 & -1/2 & 1 & -1/2 & \cdots & & & & \\ 0 & 0 & -1/2 & 1 & \cdots & & & & \\ \cdots & \cdots & \cdots & \cdots & \cdots & \cdots & \cdots & \cdots & \cdots \\ & & & & \cdots & 1 & -1/2 & 0 & 0 \\ & & 0 & & \cdots & -1/2 & 1 & -1/2 & 0 \\ & & & & \cdots & 0 & -1/2 & 1 & -1/2 \\ & & & & \cdots & 0 & 0 & -1/2 & 1 \end{array}\right].$$

The kth principal minor of A_n is just $\det A_k$, so to prove that all A_n are positive definite it will suffice to prove that $\det A_n > 0$ for all n.

Note that

$$\det A_1 = 1 = 2/2,$$
$$\det A_2 = 3/2^2,$$
$$\det A_3 = 4/2^3.$$

For $n \geq 3$ expand $\det A_n$ about the entries of the nth row. We see that

$$\det A_n = \det A_{n-1} + 1/2 \det \left[\begin{array}{c|c} A_{n-2} & 0 \\ \hline 0 \cdots 0 - 1/2 & -1/2 \end{array}\right]$$
$$= \det A_{n-1} - 1/4 \det A_{n-2}.$$

This recursion formula for $\det A_n$ can now be used to prove by induction that

$$\det A_n = (n+1)/2^n$$

for all n. Assuming the result to hold for all $k \leq n$ we have

$$\det A_{n+1} = \det A_n - 1/4 \det A_{n-1}$$
$$= (n+1)/2^n - n/2^{n+1}$$
$$= (n+2)/2^{n+1}.$$

The same procedure applies to B_n and D_n. We find that

$$\det B_n = \det B_{n-1} - 1/4 \det B_{n-2},$$
$$\det D_n = \det D_{n-1} - 1/4 \det D_{n-2},$$

and by induction that

$$\det B_n = 1/2^{n-1}, \qquad \det D_n = 1/2^{n-2}.$$

If we continue to denote $\cos \pi/5$ by α as in Exercise 2.22, then

$$I_3 = \begin{bmatrix} 1 & -\alpha & 0 \\ -\alpha & 1 & -1/2 \\ 0 & -1/2 & 1 \end{bmatrix},$$

and the principal minors of I_3 are

$$1, \qquad 1 - \alpha^2, \qquad 3/4 - \alpha^2.$$

Since $3/4 - \alpha^2 = (3 - \sqrt{5})/8 > 0$, I_3 is positive definite. The first three principal minors of I_4 are those of I_3, and

$$\begin{aligned} \det I_4 &= 1/2 - 3\alpha^2/4 \\ &= (7 - 3\sqrt{5})/32 > 0, \end{aligned}$$

so I_4 is also positive definite.

If the nodes of F_4 are labeled

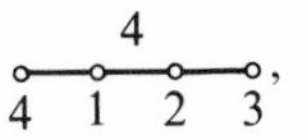

4
o—o—o—o,
4 1 2 3

then F_4 has the matrix

$$\left[\begin{array}{ccc|c} & & & -1/2 \\ & B_3 & & 0 \\ & & & 0 \\ \hline -1/2 & 0 & 0 & 1 \end{array}\right];$$

so the first three principal minors are those of B_3. Also,

$$\det F_4 = \det B_3 - 1/4 \det A_2 = 1/2^4 > 0,$$

so F_4 is positive definite.

Finally, if the nodes of E_n, $n = 6, 7, 8$, are labeled

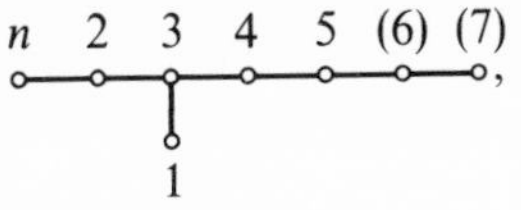

n 2 3 4 5 (6) (7)
o—o—o—o—o—o—o,
1 (attached below node 3)

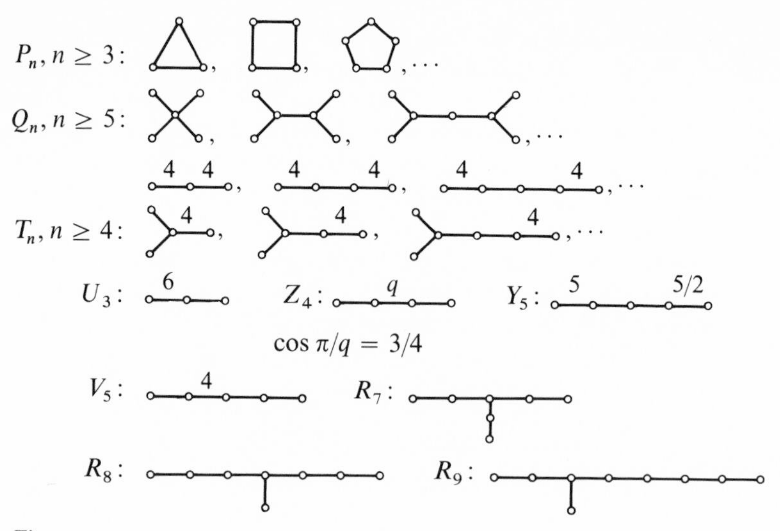

Figure 5.4

then the first $n - 1$ principal minors of E_n are those of D_{n-1}, and

$$\begin{aligned} \det E_n &= \det D_{n-1} - 1/4 \det A_{n-2} \\ &= 1/2^{n-3} - (n-1)/2^n \\ &= (8 - n + 1)/2^n > 0. \end{aligned}$$

Thus each E_n is positive definite.

We shall show presently that the list in Figure 5.3 contains *all* the connected positive definite Coxeter graphs. To this end it is convenient to consider next an auxiliary list of marked graphs. The matrix of each graph in Figure 5.4 will be shown to have zero determinant, so that none of them are positive definite. Note that $4 < q < 5$, since $\cos \pi/q = 3/4$.

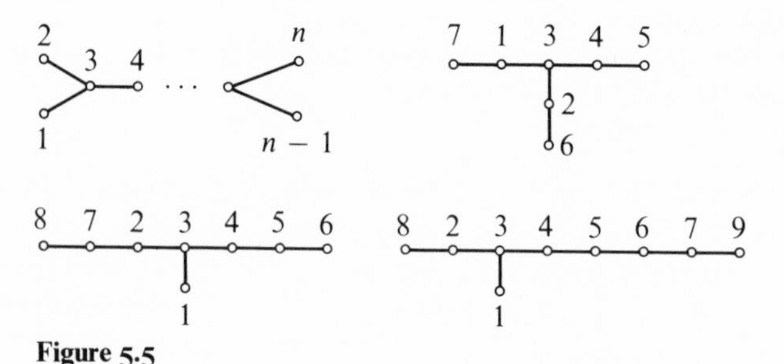

Figure 5.5

There are obvious labelings of the nodes for the graphs P_n, S_n, T_n, V_5, and Y_5. We shall use the labelings indicated in Figure 5.5 for Q_n, R_7, R_8, and R_9. The fact that $\det U_3 = \det Z_4 = 0$ will be left as an exercise.

In the case of P_n it is most convenient to observe that if the second through the nth rows are added successively to the first row, then the resulting matrix has all zero entries in the first row, and so $\det P_n = 0$.

For the other graphs in Figure 5.4 the determinant can be expanded by the minors of the last row, yielding a recursion-type relation as in the discussion of the graphs in Figure 5.3. If the nodes are labeled as indicated above, we obtain the following relations (see Exercise 2.22 for Y_5):

$$\begin{aligned}
\det Q_n &= \det D_{n-1} - 1/4 \det D_{n-3} = 0,\\
\det S_n &= \det B_{n-1} - 1/2 \det B_{n-2} = 0,\\
\det T_n &= \det D_{n-1} - 1/2 \det D_{n-2} = 0,\\
\det Y_5 &= \det I_4 - \beta^2 \det I_3\\
&= 1/2 - 3\alpha^2/4 - \beta^2(3/4 - \alpha^2)\\
&= 3(1 - 2\alpha + 2\beta)/16 = 0,\\
\det V_5 &= \det F_4 - 1/4 \det B_3 = 0,\\
\det R_7 &= \det E_6 - 1/4 \det A_5 = 0,\\
\det R_8 &= \det E_7 - 1/4 \det D_6 = 0,\\
\det R_9 &= \det E_8 - 1/4 \det E_7 = 0.
\end{aligned}$$

A marked graph H will be called a *subgraph* of a marked graph G if H can be obtained from G by removing some of the nodes of G (and any adjoining branches), or by decreasing the marks on some of the branches of G, or both. A *cycle* in a marked graph G is a subgraph of the form P_n (Figure 5.4) for some $n \geq 3$. A *branch point* in G is a node having three or more branches emanating from it. For example, each T_n has exactly one branch point.

Proposition 5.1.5

A (nonempty) subgraph H of a positive definite marked graph G is also positive definite.

Proof

Order the nodes $a_1, a_2, \ldots, a_m$ of G in such a way that $a_1, \ldots, a_k$ are the nodes of H. If $A = (\alpha_{ij})$ and $B = (\beta_{ij})$ are the matrices of G and H, respectively, then $\alpha_{ij} \leq \beta_{ij}$ for all i and j between 1 and k, since H is a subgraph of G. Let Q_G and Q_H denote the corresponding quadratic forms. If Q_H is not positive definite, choose $x \neq 0$ in $\mathscr{R}^k$ for which $Q_H(x) \leq 0$.

If $x = (\lambda_1, \ldots, \lambda_k)$, set $y = (|\lambda_1|, \ldots, |\lambda_k|, 0, \ldots, 0) \in \mathscr{R}^m$. Then $y \neq 0$, and

$$\begin{aligned} 0 \geq Q_H(x) &= \Sigma_{i,j}\, \beta_{ij}\lambda_i\lambda_j \\ &\geq \Sigma_{i,j}\, \beta_{ij}|\lambda_i|\,|\lambda_j| \\ &\geq \Sigma_{i,j}\, \alpha_{ij}|\lambda_i|\,|\lambda_j| = Q_G(y) > 0, \end{aligned}$$

a contradiction.

The importance of Proposition 5.1.5 lies in the fact that it shows that none of the graphs in Figure 5.4 can occur as subgraphs of positive definite Coxeter graphs.

Theorem 5.1.6

If G is a connected positive definite Coxeter graph, then G is one of the graphs $A_n, B_n, D_n, H_2^n, G_2, I_3, I_4, F_4, E_6, E_7$, or E_8.

Proof

Observe first that G can have no cycles as subgraphs since no P_n is positive definite. If H_2^n is a subgraph of G for any $n \geq 7$, then $G = H_2^n$, for otherwise U_3 would be a subgraph of G. Likewise, $G = G_2$ if G_2 is a subgraph. We may assume, then, for the remainder of the proof, that any branch of G is marked 3, 4, or 5. Suppose that B_2 is a subgraph of G (it cannot occur more than once; otherwise some S_n would be a subgraph). Then G cannot have a branch point, for otherwise some T_n would be a subgraph. If H_2^5 is also a subgraph, then we may have $G = H_2^5$, $G = I_3$, or $G = I_4$. There are no other possibilities in that case, for otherwise G would have either Z_4 or Y_5 as a subgraph. If B_2 is a subgraph but H_2^5 is not, then G may be B_n, for some $n \geq 2$, or F_4. There are no other possibilities, for otherwise V_5 would be a subgraph of G.

Finally, consider the case where all branches of G are unmarked. Then G can have at most one branch point, and only three branches can emanate from any branch point, for otherwise some Q_n would be a subgraph of G. If G has no branch point, then $G = A_n$ for some n. If G has a branch point, then either $G = D_n$ for some n, or $G = E_6, E_7$, or E_8; for under any other circumstance R_7, R_8, or R_9 would be a subgraph. The proof is complete.

As a consequence of Theorems 5.1.3 and 5.1.6, the Coxeter graph of an irreducible Coxeter group must appear in Figure 5.3. We shall show in Section 5.3 that, conversely, every graph in Figure 5.3 is the graph of a Coxeter group.

5.2 THE CRYSTALLOGRAPHIC CONDITION

A *lattice* in V is a set consisting of all integer linear combinations of the elements of some basis $\{x_1, \ldots, x_n\}$ for V. As in Section 2.6, a subgroup $\mathcal{G}$ of $\mathcal{O}(V)$ is said to satisfy the *crystallographic condition*, or to be a *crystallographic group*, if and only if there is a lattice $\mathcal{L}$ invariant under all transformations in $\mathcal{G}$.

We wish to determine the crystallographic Coxeter groups, and to that end it becomes convenient to allow the relative lengths of roots to differ. Thus we no longer assume that roots are unit vectors, and appropriate lengths will be assigned in the course of development. It is important to bear in mind that the formula for the reflection along a root r has the form

$$S_r x = x - \frac{2(x, r)r}{(r, r)}$$

and that the factor $1/(r, r)$ cannot in general be omitted.

Suppose as usual that $\mathcal{G}$ is a Coxeter group, with base $\Pi = \{r_1, \ldots, r_n\}$ and fundamental reflections $S_1, \ldots, S_n$. The order of $S_i S_j$ will be denoted by p_{ij} as in Proposition 5.1.1.

Proposition 5.2.1

If $\mathcal{G}$ is crystallographic, then each p_{ij} is one of the integers 1, 2, 3, 4, or 6.

Proof

Fix i and j and set $p_{ij} = m$. With respect to a suitable basis, $S_i S_j$ is represented by the matrix

$$\left[\begin{array}{c|c} A & 0 \\ \hline 0 & I_{n-2} \end{array}\right],$$

where

$$A = \begin{bmatrix} \cos 2\pi/m & -\sin 2\pi/m \\ \sin 2\pi/m & \cos 2\pi/m \end{bmatrix}.$$

Thus the trace of $S_i S_j$ is $2 \cos 2\pi/m + (n - 2)$. With respect to a lattice basis, the matrix representing $S_i S_j$ has integer entries, so $\operatorname{tr}(S_i S_j)$ must be an integer. Thus $2 \cos 2\pi/m$ is an integer. This condition is satisfied when $m = 1, 2, 3, 4$, or 6, and is not satisfied when $m = 5$ (see Exercise 2.22).

If $m > 6$, then

$$2 > 2\cos 2\pi/m > 2\cos 2\pi/6 = 1,$$

and so $2\cos 2\pi/m$ is not an integer.

As a result of Proposition 5.2.1 the only irreducible Coxeter groups that might be crystallographic are those having the Coxeter graphs $A_n, B_n, D_n, G_2, F_4, E_6, E_7$, and E_8 (provided such groups exist). Assuming their existence, which will be established in Section 5.3, let us show that these groups are in fact all crystallographic.

Suppose then that $\mathcal{G}$ is an irreducible Coxeter group and that $p_{ij} \in \{1, 2, 3, 4, 6\}$ for all i and j. Assign relative lengths to the roots $r_1, \ldots, r_n$ as follows:

If $p_{ij} = 3$, then $\|r_i\| = \|r_j\|$.
If $p_{ij} = 4$, then $\|r_i\| = \sqrt{2}\|r_j\|$ or $\|r_j\| = \sqrt{2}\|r_i\|$.
If $p_{ij} = 6$, then $\|r_i\| = \sqrt{3}\|r_j\|$ or $\|r_j\| = \sqrt{3}\|r_i\|$.

It is easy to see by inspection of the Coxeter graphs that it is always possible to assign lengths consistently that satisfy these requirements. All simple roots must be taken to have equal lengths in the cases A_n, D_n, E_6, E_7, and E_8. For B_n we may take

$$\sqrt{2}\|r_1\| = \|r_2\| = \cdots = \|r_n\|,$$

for G_2 take $\sqrt{3}\|r_1\| = \|r_2\|$, and for F_4 take

$$\|r_1\| = \|r_2\| = \sqrt{2}\|r_3\| = \sqrt{2}\|r_4\|.$$

Theorem 5.2.2

If $\mathcal{G}$ has Coxeter graph $A_n, B_n, D_n, G_2, F_4, E_6, E_7$, or E_8, then $\mathcal{G}$ satisfies the crystallographic condition.

Proof

Let $\mathcal{L}$ be the lattice having Π as basis,

$$\mathcal{L} = \{\Sigma_{i=1}^{n} k_i r_i : k_i \in \mathbb{Z}\}.$$

If $p_{ij} = 3$, then $\|r_i\| = \|r_j\|$ and

$$(r_i, r_j) = (-1/2)\|r_i\|\ \|r_j\|.$$

Thus

$$S_i r_j = r_j - \frac{2(r_j, r_i) r_i}{(r_i, r_i)} = r_i + r_j \in \mathcal{L}.$$

Suppose $p_{ij} = 4$, so that

$$(r_i, r_j) = -(\sqrt{2}/2)\|r_i\|\ \|r_j\|.$$

If $\|r_i\| = \sqrt{2}\|r_j\|$, then $(r_i, r_j) = -\|r_j\|^2$ and

$$S_i r_j = r_j - \frac{2(-\|r_j\|^2)}{2\|r_j\|^2} r_i = r_j + r_i \in \mathscr{L}.$$

If $\|r_j\| = \sqrt{2}\|r_i\|$, then $(r_i, r_j) = -\|r_i\|^2$ and

$$S_i r_j = r_j + 2r_i \in \mathscr{L}.$$

A similar analysis shows that if $p_{ij} = 6$, then $S_i r_j$ must be either $r_j + r_i$ or $r_j + 3r_i$, so $S_i r_j \in \mathscr{L}$. Of course, if $p_{ij} = 1$, then $S_i r_j = -r_j$; and if $p_{ij} = 2$, then $S_i r_j = r_j$. It follows that $S_i\mathscr{L} = \mathscr{L}$ for every fundamental reflection S_i, and hence that $T\mathscr{L} = \mathscr{L}$ for all $T \in \mathscr{G}$, by Theorem 4.1.12. Thus $\mathscr{G}$ is crystallographic.

For reducible Coxeter groups $\mathscr{G}$ see Exercise 5.3.

4.3 CONSTRUCTION OF IRREDUCIBLE COXETER GROUPS

Two different methods will be used to establish the existence of the various irreducible Coxeter groups. The first method, which will be applied to the groups with graphs A_n, B_n, and D_n, involves finding a group "in nature," so to speak, and then proving that it is a reflection group. The second method proceeds directly with a construction of the group from reflections along a set of vectors that will ultimately be a base for the root system.

The dihedral groups, their root systems, and bases have been discussed earlier (see p. 40 and Exercise 4.3).

We may view the symmetric group $\mathscr{S}_{n+1}$ as a group of linear transformations of $\mathscr{R}^{n+1}$, if we agree that each $T \in \mathscr{S}_{n+1}$ is a permutation of the basis vectors $e_1, e_2, \ldots, e_{n+1}$. It is well known that $\mathscr{S}_{n+1}$ is generated by the n successive transpositions

$$S_1 = (e_1 e_2), S_2 = (e_2 e_3), \ldots, S_n = (e_n e_{n+1}).$$

If the permutations S_i are considered as linear transformations, it is immediate that they satisfy

$$S_i(e_{i+1} - e_i) = -(e_{i+1} - e_i),$$

$$S_i(e_i + e_{i+1}) = e_i + e_{i+1},$$

$$S_i(e_j) = e_j \qquad \text{if } j \neq i, j \neq i+1.$$

Since $\{e_j : j \neq i, j \neq i+1\} \cup \{e_i + e_{i+1}\}$ spans the hyperplane $W = (e_{i+1} - e_i)^{\perp}$, we see that each S_i is a reflection and that $r_i = e_{i+1} - e_i$ may be taken as a root of S_i. Thus $\mathscr{S}_{n+1}$, as a group of transformations, is generated by the reflections $S_1, \ldots, S_n$ along roots $r_1, \ldots, r_n$. In particular, $\mathscr{S}_{n+1} \leq \mathcal{O}(\mathscr{R}^{n+1})$.

To obtain the root system Δ of $\mathscr{S}_{n+1}$ we must find the roots of all conjugates in $\mathscr{S}_{n+1}$ of the generating reflections S_i. But in a symmetric group the set of conjugates of any transposition is the set of *all* transpositions. Thus the set of conjugate reflections is the set of transpositions $(e_i e_j)$, $i \neq j$, and the root system is

$$\Delta = \{e_i - e_j : i \neq j, 1 \leq i, j \leq n+1\}.$$

Let us denote by V the subspace of $\mathscr{R}^{n+1}$ spanned by the roots $r_1, \ldots, r_n$ of $\mathscr{S}_{n+1}$, and by $\mathscr{A}_n$ the group of restrictions to V of the transformations in $\mathscr{S}_{n+1}$. ($\mathscr{A}_n$ is *not* the alternating group.) We continue to denote the generating reflections by $S_1, \ldots, S_n$. Then $\mathscr{A}_n$ is effective, and hence is a Coxeter group. Since $\mathscr{A}_n$ is isomorphic with $\mathscr{S}_{n+1}$, we have $|\mathscr{A}_n| = (n+1)!$.

Observe that if $x = (\lambda_1, \ldots, \lambda_{n+1}) \in \mathscr{R}^{n+1}$, then $x \in V^{\perp}$ if and only if $\lambda_{i+1} - \lambda_i = 0$, $1 \leq i \leq n$, since $x \perp r_i$; so $x = \lambda_1(1, 1, \ldots, 1)$. As a result,

$$V = \{y \in \mathscr{R}^{n+1} : y = (\mu_1, \ldots, \mu_{n+1}), \Sigma\, \mu_i = 0\}$$

If $r = e_i - e_j \in \Delta$, with $i > j$, then

$$r = \Sigma_{k=j}^{i-1}(e_{k+1} - e_k) = \Sigma_{k=j}^{i-1} r_k.$$

Choose $t \in V$ such that $(t, r_i) > 0$, $1 \leq i \leq n$ (see Exercise 5.4, or simply compute). Then $\{r_1, \ldots, r_n\}$ is a t-base for $\mathscr{A}_n$, and

$$\Delta^+ = \{r \in \Delta : r = e_i - e_j, i > j\},$$
$$\Delta^- = \{r \in \Delta : r = e_i - e_j, i < j\}.$$

Finally, since

$$(r_i, r_{i+1})/\|r_i\|\,\|r_{i+1}\| = -1/2,$$

$1 \leq i \leq n-1$, the Coxeter graph of $\mathscr{A}_n$ is A_n.

The discussion of groups with Coxeter graphs B_n and D_n is similar but slightly more complicated.

For B_n we consider the group of "signed permutations" in $\mathscr{R}^n$ whose elements permute the basis vectors $e_1, \ldots, e_n$, and then replace some of them by their negatives. More precisely, we begin by considering two subgroups of $\mathcal{O}(\mathscr{R}^n)$. The first is the symmetric group $\mathscr{S}_n$ of permutations of $\{e_1, \ldots, e_n\}$. Changing the notation slightly from the discussion of

$\mathscr{A}_n$ above, we denote the generating reflections of $\mathscr{S}_n$ by $S_2, S_3, \ldots, S_n$, with respective roots $r_2 = e_2 - e_1, \ldots, r_n = e_n - e_{n-1}$. The second subgroup $\mathscr{K}_n$ is generated by the n reflections $S_{e_1}, \ldots, S_{e_n}$ along the basis vectors $e_1, \ldots, e_n$. The effect of S_{e_i} on any $x \in \mathscr{R}^n$ is simply to replace the ith entry of x by its negative. Since the reflections S_{e_i} all commute with one another, $\mathscr{K}_n$ is abelian and is the direct product of the two-element subgroups $\{1, S_{e_i}\}$, $1 \leq i \leq n$. Thus $|\mathscr{K}_n| = 2^n$.

For each subset J of $\{e_1, \ldots, e_n\}$, define $f_J : \mathscr{R}^n \to \mathscr{R}^n$ by setting

$$f_J(e_i) = \begin{cases} -e_i & \text{if } e_i \in J, \\ e_i & \text{if } e_i \notin J. \end{cases}$$

Clearly, f_J is the product of all S_{e_i} for which $e_i \in J$, with $f_J = 1$ if $J = \varnothing$, and $\mathscr{K}_n$ consists of all such transformations f_J. The *symmetric difference* $J \dotplus L$ of two sets J and L is defined by

$$J \dotplus L = (J \cup L) \setminus (J \cap L).$$

It is easy to check that if $f_J, f_L \in \mathscr{K}_n$, then $f_J f_L = f_{J \dotplus L}$.

Let us investigate how the subgroups $\mathscr{S}_n$ and $\mathscr{K}_n$ interact. Suppose that $T \in \mathscr{S}_n$ and $f_J \in \mathscr{K}_n$. Given a basis vector e_i, write $e_i = Te_j$. If $e_j \in J$ or, equivalently, if $e_i \in T(J)$, then

$$\begin{aligned}(Tf_J T^{-1})e_i &= (Tf_J T^{-1})Te_j \\ &= Tf_J e_j = -Te_j = -e_i;\end{aligned}$$

and if $e_j \notin J$, or $e_i \notin T(J)$, then

$$(Tf_J T^{-1})e_i = Tf_J e_j = Te_j = e_i.$$

Thus

$$Tf_J T^{-1} = f_{T(J)},$$

and $\mathscr{K}_n$ is *normalized* by $\mathscr{S}_n$. It follows that the subgroup $\mathscr{B}_n$ of $\mathscr{O}(\mathscr{R}^n)$ that is generated by $\mathscr{K}_n \cup \mathscr{S}_n$ consists simply of all products $f_J T$, where $f_J \in \mathscr{K}_n$ and $T \in \mathscr{S}_n$. By considering their effects on the basis vectors $e_1, \ldots, e_n$, we see that $\mathscr{K}_n \cap \mathscr{S}_n = 1$, from which it follows that each transformation in $\mathscr{B}_n$ has a *unique* expression as a product $f_J T$. Thus the order of $\mathscr{B}_n$ is the product of the orders of $\mathscr{K}_n$ and $\mathscr{S}_n$; i.e.,

$$|\mathscr{B}_n| = |\mathscr{K}_n|\,|\mathscr{S}_n| = 2^n \cdot n!. \quad .$$

Products in $\mathscr{B}_n$ are given by

$$\begin{aligned}(f_J T)(f_L U) &= f_J T f_L T^{-1} T U \\ &= f_J f_{T(L)} T U = f_{J \dotplus T(L)} \cdot TU.\end{aligned}$$

Readers having some familiarity with group extensions will recognize that $\mathscr{B}_n$ is a *semidirect product*, or *split extension*, of $\mathscr{K}_n$ by $\mathscr{S}_n$.

If we set $f_i = f_{\{e_i\}} = S_{e_i}$, $1 \le i \le n$, then $\mathscr{B}_n$ is generated by the set of reflections

$$\{f_1, \ldots, f_n, S_2, \ldots, S_n\}.$$

Among the roots of $\mathscr{B}_n$ are those of $\mathscr{S}_n$. In particular, $e_i - e_1$ is a root if $2 \le i \le n$. Let T_i denote, momentarily, the reflection with root $e_i - e_1$, and observe that

$$T_i f_1 T_i^{-1} = f_{T_i\{e_1\}} = f_{\{e_i\}} = f_i.$$

Consequently, $\{f_1, S_2, \ldots, S_n\}$ is a generating set of reflections for $\mathscr{B}_n$. Set $S_1 = f_1$ and set $r_1 = e_1$, a root of S_1. Since $\{r_1, \ldots, r_n\}$ is a basis for $\mathscr{R}^n$, $\mathscr{B}_n$ is a Coxeter group.

Since the effect of each transformation in $\mathscr{B}_n$ on the basis vectors $e_1, \ldots, e_n$ is a permutation, followed by some sign changes, the root system Δ of $\mathscr{B}_n$ is

$$\Delta = \{\pm e_i : 1 \le i \le n\} \cup \{e_i \pm e_j : i \ne j, 1 \le i, j \le n\}.$$

Easy computations show that

$$\begin{aligned} e_i &= \textstyle\sum_{k=1}^{i} r_k, \\ e_i - e_j &= \textstyle\sum_{k=j}^{i-1} r_{k+1} && \text{if } i > j, \\ e_i + e_j &= 2\textstyle\sum_{k=1}^{j} r_k + \textstyle\sum_{k=j}^{i-1} r_{k+1} && \text{if } i > j; \end{aligned}$$

so $\{r_1, \ldots, r_n\}$ is a base for $\mathscr{B}_n$. The Coxeter graph of $\mathscr{B}_n$ is B_n since

$$(r_1, r_2)/\|r_1\|\,\|r_2\| = -\sqrt{2}/2.$$

Note that the roots of $\mathscr{B}_n$ were chosen to have lengths in accordance with the crystallographic condition.

For the discussion of D_n we replace $\mathscr{K}_n$ by its subgroup,

$$\mathscr{L}_n = \{f_J \in \mathscr{K}_n : |J| \text{ is even}\},$$

whose transformations effect even numbers of sign changes in coordinates of vectors in $\mathscr{R}^n$. That $\mathscr{L}_n$ is a subgroup of $\mathscr{K}_n$ is a consequence of the fact that

$$|J \dotplus L| = |J| + |L| - 2|J \cap L|$$

for any finite sets J and L.

As above, $\mathscr{L}_n$ is normalized by $\mathscr{S}_n$ and the subgroup $\mathscr{D}_n$ of $\mathscr{O}(\mathscr{R}^n)$ generated by $\mathscr{L}_n \cup \mathscr{S}_n$ consists of all products $f_J T$, $f_J \in \mathscr{L}_n$, $T \in \mathscr{S}_n$. Set $r_1 = e_1 + e_2$ and $r_i = e_i - e_{i+1}$, $2 \le i \le n$, and let $S_j \in \mathscr{D}_n$ be the reflection along r_j, $1 \le j \le n$. Observe that $\langle S_2, \ldots, S_n \rangle = \mathscr{S}_n$.

Since any even number of sign changes can be made by successively changing signs two at a time, $\mathcal{L}_n$ is generated by the products $S_{e_i}S_{e_j}$, $i \neq j$. Among the transformations in $\mathcal{D}_n$ are the reflections $S_{e_i+e_j}$, $i \neq j$. Given $i \neq j$ choose $T \in \mathcal{S}_n$ such that $Te_1 = e_i$ and $Te_2 = e_j$. Then by Proposition 4.1.1 we have

$$TS_{e_1+e_2}T^{-1} = S_{e_i+e_j}.$$

Also, if $i \neq j$, then $S_{e_i-e_j} \in \mathcal{S}_n$, and it is easy to check that

$$S_{e_i-e_j}S_{e_i+e_j} = S_{e_i}S_{e_j}.$$

It follows that $\{S_1, S_2, \ldots, S_n\}$ is a generating set of reflections for $\mathcal{D}_n$.

Applying all permutations of $\{e_i\}$ to the roots $\{r_1, \ldots, r_n\}$, followed by even numbers of sign changes, we obtain the root system

$$\Delta = \{e_i \pm e_j : i \neq j, 1 \leq i, j \leq n\}.$$

Since

$$\begin{aligned} e_i - e_j &= \Sigma_{k=j}^{i-1} r_{k+1} && \text{if } i > j, \\ e_i + e_1 &= r_1 + \Sigma_{k=3}^{i} r_k && \text{if } i \neq 1, \\ e_i + e_2 &= \Sigma_{k=1}^{i} r_k && \text{if } i > 2, \end{aligned}$$

and

$$e_i + e_j = r_1 + r_2 + 2\,\Sigma_{k=3}^{j} r_k + \Sigma_{k=j}^{i-1} r_{k+1} \qquad \text{if } 2 < j < i,$$

we see that $\{r_1, \ldots, r_n\}$ is a base for $\mathcal{D}_n$. Its Coxeter graph is D_n.

The order of $\mathcal{D}_n$ is easily found by counting, but observe that the function θ from $\mathcal{B}_n$ to $\{\pm 1\}$ defined by

$$\theta(f_J T) = (-1)^{|J|}$$

is a homomorphism onto $\{\pm 1\}$ with kernel $\mathcal{D}_n$. Thus $[\mathcal{B}_n : \mathcal{D}_n] = 2$ and

$$|\mathcal{D}_n| = 1/2|\mathcal{B}_n| = 2^{n-1} \cdot n!.$$

In order to establish the existence of a group for each of the remaining graphs G in Figure 5.3, we shall exhibit a set of mutually obtuse vectors with Coxeter graph G. Using them we shall construct a Coxeter group $\mathcal{G}$ and its root system Δ, and show that the original set of vectors is a base for $\mathcal{G}$. If $\mathcal{G}$ is crystallographic, the base will be chosen so as to generate a $\mathcal{G}$-invariant lattice.

In order to illustrate a general method for producing a base with a prescribed Coxeter graph, let us discuss the case of F_4 in some detail. The graph B_3 is a subgraph of F_4, and a base for the group $\mathcal{B}_3$ was discussed above. We shall "extend" that base to a base having the graph F_4.

If the base vectors r_1, r_2, and r_3 of $\mathscr{B}_3$ are relabeled as

$$r_2 = e_1, \qquad r_3 = e_2 - e_1, \qquad r_4 = e_3 - e_2$$

and viewed as vectors in $\mathscr{R}^4$, then we wish to add a vector r_1 so that $\{r_1, r_2, r_3, r_4\}$ has Coxeter graph F_4. The requirements for $r_1 = (\lambda_1, \lambda_2, \lambda_3, \lambda_4)$, including the crystallographic condition, are

$$\begin{aligned} \|r_1\|^2 = \|r_2\|^2 = \Sigma_{i=1}^4 \lambda_i^2 &= 1, \\ (r_1, r_2) = \lambda_1 &= -1/2, \\ (r_1, r_3) = \lambda_2 - \lambda_1 &= 0, \\ (r_1, r_4) = \lambda_3 - \lambda_2 &= 0. \end{aligned}$$

Thus $r_1 = (-1/2, -1/2, -1/2, \lambda_4)$, with $\lambda_4^2 = 1/4$. We choose $\lambda_4 = -1/2$, or

$$r_1 = -(1/2)\,\Sigma_{i=1}^4 e_i,$$

and the resulting graph is F_4.

For G_2 it is convenient to choose a base in $\mathscr{R}^3$, extending the one basic vector $r_1 = e_2 - e_1$ of $\mathscr{A}_1$. Of course, r_2 could be chosen in $\mathscr{R}^2$, but it is possible to find $r_2 \in \mathscr{R}^3$ with *integer* coordinates such that $\{r_1, r_2\}$ has graph G_2. The requirements for $r_2 = (\lambda_1, \lambda_2, \lambda_3)$ are

$$\begin{aligned} \lambda_1^2 + \lambda_2^2 + \lambda_3^2 &= 6, \\ (r_1, r_2) = \lambda_2 - \lambda_1 &= -3. \end{aligned}$$

There are many solutions, but we take

$$\lambda_1 = 1, \qquad \lambda_2 = -2, \qquad \lambda_3 = 1,$$

or $r_2 = e_1 - 2e_2 + e_3$, and $\{r_1, r_2\}$ has graph G_2.

In Section 4.2 we found a base $\{r_1, r_2, r_3\}$ of unit vectors for the icosahedral group $\mathscr{I}^*$, with

$$\begin{aligned} r_1 &= \beta(2\alpha + 1, 1, -2\alpha), \\ r_2 &= \beta(-2\alpha - 1, 1, 2\alpha), \\ r_3 &= \beta(2\alpha, -2\alpha - 1, 1). \end{aligned}$$

Using Exercise 2.22 we find that

$$\begin{aligned} (r_1, r_2) &= -\alpha = -\cos \pi/5, \\ (r_1, r_3) &= 0, \\ (r_2, r_3) &= -1/2; \end{aligned}$$

so $\{r_1, r_2, r_3\}$ has Coxeter graph I_3. By the above procedure we may view

r_1, r_2, and r_3 as vectors in $\mathscr{R}^4$ and extend to a base with graph I_4 by adjoining the vector $r_4 = \beta(-2\alpha,\ 0,\ -2\alpha - 1, 1)$.

One more extension is required. If the base vectors of $\mathscr{A}_7$ are again relabeled as

$$r_i = e_i - e_{i-1}, \qquad 2 \le i \le 8,$$

we may adjoin

$$r_1 = (1/2)(\Sigma_{i=1}^3 e_i - \Sigma_{i=4}^8 e_i);$$

then $\{r_1, r_2, \ldots, r_8\}$ has graph E_8, since

$$(r_1, r_4)/\|r_1\|\ \|r_4\| = -1/2,$$
$$(r_1, r_i) = 0 \qquad \text{if } i \neq 1, i \neq 4.$$

Clearly, the subsets $\{r_1, \ldots, r_6\}$ and $\{r_1, \ldots, r_7\}$ have graphs E_6 and E_7.

The bases for all the graphs in Figure 5.3 are tabulated in Table 5.1 for easy reference.

For each graph G from G_2 to E_8 in Table 5.1, let V be the space spanned by the set of vectors $\Gamma = \{r_1, \ldots, r_n\}$. In each case Γ is a linearly independent set so it is a basis for V. Thus there is a vector $t \in V$ such that $(t, r_i) > 0$, $1 \le i \le n$ (see Exercise 5.4). Let S_i be the reflection in $\mathcal{O}(V)$ along r_i, $1 \le i \le n$, and let

$$\mathscr{G} = \langle S_1, \ldots, S_n \rangle \le \mathcal{O}(V).$$

Graph	Base
A_n	$r_i = e_{i+1} - e_i,\ 1 \le i \le n.$
B_n	$r_1 = e_1, r_i = e_i - e_{i-1}, 2 \le i \le n.$
D_n	$r_1 = e_1 + e_2, r_i = e_i - e_{i-1}, 2 \le i \le n.$
H_2^n	$r_1 = (1, 0), r_2 = (-\cos \pi/n, \sin \pi/n).$
G_2	$r_1 = e_2 - e_1, r_2 = e_1 - 2e_2 + e_3.$
I_3	$r_1 = \beta(2\alpha + 1, 1, -2\alpha), r_2 = \beta(-2\alpha - 1, 1, 2\alpha),$ $r_3 = \beta(2\alpha, -2\alpha - 1, 1).$
I_4	$r_1 = \beta(2\alpha + 1, 1, -2\alpha, 0), r_2 = \beta(-2\alpha - 1, 1, 2\alpha, 0),$ $r_3 = \beta(2\alpha, -2\alpha - 1, 1, 0), r_4 = \beta(-2\alpha, 0, -2\alpha - 1, 1).$
F_4	$r_1 = -(1/2)\Sigma_1^4 e_i, r_2 = e_1, r_3 = e_2 - e_1, r_4 = e_3 - e_2.$
E_6	$r_1 = (1/2)(\Sigma_1^3 e_i - \Sigma_4^8 e_i), r_i = e_i - e_{i-1}, 2 \le i \le 6.$
E_7	$r_1 = (1/2)(\Sigma_1^3 e_i - \Sigma_4^8 e_i), r_i = e_i - e_{i-1}, 2 \le i \le 7.$
E_8	$r_1 = (1/2)(\Sigma_1^3 e_i - \Sigma_4^8 e_i), r_i = e_i - e_{i-1}, 2 \le i \le 8.$

Table 5.1

The following algorithm can be used to obtain the root system Δ of $\mathscr{G}$. Set $\Gamma_0 = \Gamma = \{r_1, \ldots, r_n\}$. For each $r_i \in \Gamma_0$ with $(r_1, r_i) < 0$, apply S_1 to r_i to obtain the root $S_1 r_i$. The same procedure is applied in turn with S_1 and r_1 replaced by S_2 and r_2, then by S_3 and $r_3, \ldots, S_n$ and r_n. Denote the set of roots obtained, including Γ_0, by Γ_1. The process is then repeated, with S_i applied to each $r \in \Gamma_1 \setminus \Gamma_0$ for which $(r_i, r) < 0$, $1 \leq i \leq n$, and the resulting set of roots is denoted by Γ_2. Continuing, we obtain

$$\Gamma_0 \subseteq \Gamma_1 \subseteq \Gamma_2 \subseteq \Gamma_3 \subseteq \cdots.$$

Since S_i is applied to $r \in \Gamma_j$ only if $(r_i, r) < 0$, and since

$$S_i r = r - \frac{2(r_i, r)}{(r_i, r_i)} r_i,$$

we see inductively that every root obtained by this process is a nonnegative linear combination of $\{r_1, \ldots, r_n\}$.

When the algorithm is applied to the bases in Table 5.1, we find that in each case the procedure terminates in a finite number of steps, in the sense that for some k we have $(r_i, r) \geq 0$ for all $r \in \Gamma_k \setminus \Gamma_{k-1}$, $1 \leq i \leq n$. We then adjoin the negatives of all roots obtained, setting $\Gamma^* = \Gamma_k \cup -\Gamma_k$.

The next step is to verify that $S_i \Gamma^* = \Gamma^*$, $1 \leq i \leq n$, and hence that $T\Gamma^* = \Gamma^*$ for all $T \in \mathscr{G}$ since $\mathscr{G}$ is generated by $\{S_1, \ldots, S_n\}$. It follows that $\Gamma^* = \Delta$, the root system of $\mathscr{G}$, by the definition of Δ. Note that it is not a priori clear that the algorithm has produced the set of *all* roots of $\mathscr{G}$. This is the reason that Δ was defined as it was in Section 4.1.

Since Δ contains a basis for V, the group $\mathscr{G}$ is effective; and since Δ is finite in each case, $\mathscr{G}$ is a Coxeter group, by Proposition 4.1.3. Since the algorithm produced only roots that are nonnegative linear combinations of $\{r_1, \ldots, r_n\}$, and since the only other roots are negatives of those, we see that $\{r_1, \ldots, r_n\}$ is a t-base for Δ. Thus G is the graph of the Coxeter group $\mathscr{G}$, as we wished to show.

Let us apply the algorithm in the case of G_2, the simplest case, even though there is no question of the existence of a group with graph G_2. Thus

$$\Gamma_0 = \{r_1, r_2\} = \{e_2 - e_1, e_1 - 2e_2 + e_3\},$$

and we set $\mathscr{G}_2 = \langle S_1, S_2 \rangle$. Then $(r_1, r_2) = -3 < 0$, so at the first stage we obtain

$$S_1 r_2 = 3r_1 + r_2, \qquad S_2 r_1 = r_1 + r_2,$$
$$\Gamma_1 \setminus \Gamma_0 = \{r_1 + r_2, 3r_1 + r_2\}$$
$$= \{e_3 - e_2, -2e_1 + e_2 + e_3\}.$$

At the next stage

$$(r_1, r_1 + r_2) = 2 - 3 < 0, \qquad (r_1, 3r_1 + r_2) = 6 - 3 > 0,$$
$$(r_2, r_1 + r_2) = -3 + 6 > 0, \qquad (r_2, 3r_1 + r_2) = -9 + 6 < 0,$$

and we obtain

$$\begin{aligned} S_1(r_1 + r_2) &= -r_1 + (3r_1 + r_2) = 2r_1 + r_2, \\ S_2(3r_1 + r_2) &= 3(r_1 + r_2) - r_2 = 3r_1 + 2r_2, \\ \Gamma_2 \setminus \Gamma_1 &= \{2r_1 + r_2, 3r_1 + 2r_2\} \\ &= \{e_3 - e_1, -e_1 - e_2 + 2e_3\}. \end{aligned}$$

The procedure terminates at this stage, since

$$(r_1, 2r_1 + r_2) = 4 - 3 > 0, \qquad (r_1, 3r_1 + 2r_2) = 6 - 6 = 0,$$
$$(r_2, 2r_1 + r_2) = -6 + 6 = 0, \qquad (r_2, 3r_1 + 2r_2) = -9 + 12 > 0.$$

Thus $\Gamma^* = \Gamma_2 \cup -\Gamma_2$.

In order to verify that $S_i\Gamma^* = \Gamma^*$ it suffices to check that $S_i\Gamma_2 \subseteq \Gamma^*$; if $r \in \Gamma_2$, we know already that $S_i r \in \Gamma_2$ if $(r_i, r) < 0$. Of course, $S_i r = r$ if $(r_i, r) = 0$, so it remains only to verify that $S_i r \in \Gamma^*$ if $r \in \Gamma_2$ and $(r_i, r) > 0$. If $r = r_i$, then $S_i r = -r \in \Gamma^*$, so the only cases left are

$$i = 1: \qquad r = 3r_1 + r_2, 2r_1 + r_2;$$
$$i = 2: \qquad r = r_1 + r_2, 3r_1 + 2r_2.$$

But in each of these cases $r = S_i r'$ for some $r' \in \Gamma_2$ [e.g., $2r_1 + r_2 = S_1(r_1 + r_2)$], so

$$S_i r = S_i^2 r' = r' \in \Gamma_2.$$

As discussed above, it follows that $\Delta = \Gamma^*$ and that $\mathscr{G}_2$ is a Coxeter group with base $\{r_1, r_2\}$ and Coxeter graph G_2.

The algorithm is easily carried out for $G = I_3$ and $G = F_4$. We omit the details and list the groups and their root systems in Table 5.2.

The construction can be simplified considerably for $G = E_8$ if we modify the algorithm. As usual we set $\mathscr{E}_8 = \langle S_1, \ldots, S_8 \rangle \leq \mathscr{O}(V)$. Since $\{r_2, \ldots, r_8\}$ is a base for $\mathscr{A}_7$, $\mathscr{A}_7$ is a subgroup of $\mathscr{E}_8$, and among the roots of $\mathscr{E}_8$ are the 28 positive roots $r_{ij} = e_i - e_j$, $1 \leq j < i \leq 8$, of $\mathscr{A}_7$. We set $S_{ij} = S_{r_{ij}}$ and begin the algorithm with the larger set

$$\Gamma_0 = \{r_{ij} : 1 \leq j < i \leq 8\} \cup \{r_1\}.$$

If $j \leq 3 < i$, then $(r_1, r_{ij}) = -1$ and $S_1 r_{ij} = r_{ij} + r_1$ is a root, the effect being a change of sign in the ith and jth entries of r_1. At the next stage we have, for example, $(r_{52}, r_{41} + r_1) = -1$; so $r_1 + r_{41} + r_{52}$

is a root, changing the first, second, fourth and fifth signs in r_1. Continuing, we obtain as roots all vectors that result from permuting the entries of r_1. In particular, we obtain

$$r = (1/2)(-1, -1, -1, 1, 1, 1, -1, -1),$$

and $(r, r_1) = -1$, so

$$S_1 r = r + r_1 = -e_7 - e_8$$

is a root.

If $i \leq 6$, then $(r_{7i}, -e_7 - e_8) = -1$, and we obtain

$$S_{7i} S_1 r = (-e_7 - e_8) + (e_7 - e_i) = -e_i - e_8 .$$

Then if $j \neq i$, $j \neq 8$, we also have $(r_{8j}, -e_i - e_8) = -1$, and we obtain all $-e_i - e_j, i \neq j$.

Observe next, for example, that $(r_1, -e_2 - e_3) = -1$, so that $r_1 - e_2 - e_3 = (1/2)(1, -1, -1, -1, -1, -1, -1, -1)$ is a root. Similarly, we obtain all vectors $(1/2) \Sigma_1^8 \varepsilon_i e_i$, where one ε_i is 1 and the others are -1.

All the roots of $\mathscr{E}_8$ obtained thus far are nonnegative linear combinations of $\{r_1, r_2, \ldots, r_8\}$ (having, in fact, integer coefficients). Adjoining their negatives we have the following set Γ^* of 240 roots:

$$\begin{array}{lll} \text{all } \pm e_i \pm e_j, & i \neq j, & 1 \leq i, j \leq 8, \\ \text{all } (1/2) \Sigma_1^8 \varepsilon_i e_i, & \varepsilon_i = \pm 1, & \Pi_{i=1}^8 \varepsilon_i = -1. \end{array}$$

In order to show that $\Gamma^* = \Delta$, the root system of $\mathscr{E}_8$, it will suffice to show that $S_i \Gamma^* = \Gamma^*$, $1 \leq i \leq 8$. The roots $\pm e_i \pm e_j$ are the roots of $\mathscr{D}_8$, as constructed above, and $\mathscr{D}_8$ has $S_2, \ldots, S_8$ among its reflections, so those roots are invariant under $S_2, \ldots, S_8$. As for the effect of S_1 on those vectors, it will suffice to consider $S_1(e_i \pm e_j)$, with $i > j$. Let us momentarily call roots of the form $(1/2) \Sigma \varepsilon_i e_i$ *1/2-vectors*. It is easily verified that

$$S_1(e_i + e_j) = e_i + e_j$$

if $j \leq 3 < i$; that

$$S_1(e_i + e_j) = e_i + e_j - r_1$$

is a 1/2-vector with one minus sign if $i \leq 3$; that

$$S_1(e_i + e_j) = e_i + e_j + r_1$$

is a 1/2-vector with three minus signs if $j > 3$; that

$$S_1(e_i - e_j) = e_i - e_j$$

if $i \leq 3$ or $j \geq 4$; and that

$$S_1(e_i - e_j) = e_i - e_j + r_1$$

is a 1/2-vector with three minus signs if $j \leq 3 < i$.

It remains to check the effects of all S_i on the 1/2-vectors $(1/2)\Sigma\, \varepsilon_i e_i$. If $i \neq 1$ and r is a 1/2-vector, then either $S_i r = r$ or else $S_i r$ is another 1/2-vector with two sign changes. If we write

$$r_1 = (1/2)\Sigma\, v_i e_i, \qquad r = (1/2)\Sigma\, \varepsilon_i e_i,$$

then $(r_1, r) = (1/4)\Sigma\, v_i \varepsilon_i$ and

$$\Pi_1^8(v_i\varepsilon_i) = (\Pi v_i)(\Pi \varepsilon_i) = (-1)^2 = 1;$$

so $v_i \varepsilon_i = -1$ for an even number of indices i. Assuming that $r \neq \pm r_1$, the number is two, four, or six. If it is four, then $(r_1, r) = 0$ and $S_1 r = r$. If it is two, then $(r_1, r) = 1$ and $S_1 r = r - r_1$. But then r_1 and r agree except at two entries, so $r - r_1$ is $\pm e_i \pm e_j$ for some i and j. Similarly, if $v_i \varepsilon_i = -1$ for six values of i, then $S_1 r = r + r_1$, and again $S_1 r$ is $\pm e_i \pm e_j$ for some i and j.

Thus $\Gamma^* = \Delta$, and we may conclude as in the earlier cases that $\mathscr{E}_8$ is a Coxeter group with Coxeter graph E_8.

If we set $\mathscr{E}_7 = \langle S_1, \ldots, S_7 \rangle$, then $\mathscr{E}_7 \leq \mathscr{E}_8$, and the roots of $\mathscr{E}_7$ are among those of $\mathscr{E}_8$. The root $u = (1/2)(1, 1, 1, 1, 1, 1, 1, -1)$ of $\mathscr{E}_8$ is orthogonal to $r_1, \ldots, r_7$, so the roots of $\mathscr{E}_7$ are precisely the roots of $\mathscr{E}_8$ that are orthogonal to u. These are the 48 roots $e_i - e_j$, $i, j \neq 8$; the 14 roots $\pm(e_i + e_8)$, $i \neq 8$; and the 70 1/2-vectors having either three negative entries with the eighth positive or five negative entries including the eighth. Thus $|\Delta| = 126$ for $\mathscr{E}_7$.

A similar discussion shows that the roots of $\mathscr{E}_6 = \langle S_1, \ldots, S_6 \rangle$ are the 72 roots of $\mathscr{E}_7$ that are also orthogonal to $r_8 = e_8 - e_7$.

The only group remaining is $\mathscr{I}_4$, with graph I_4. As in the case of $\mathscr{E}_8$, we may modify the algorithm by including at the outset the 15 positive roots of $\mathscr{I}_3$. Details will be omitted, but the reader should be warned that the computations are still rather tedious, the chief reason being that several applications of the identities that appear in Exercise 2.22 are required. The algorithm is readily programmed for computer calculation.

If we agree that

$$(\lambda_1, \lambda_2, \lambda_3, \lambda_4) = \lambda_1 + \lambda_2 i + \lambda_3 j + \lambda_4 k,$$

then $\mathscr{R}^4$ can be identified with the ring Q of real quaternions (see [19], p. 87, or [1], p. 222; also see Exercise 5.14). It is a remarkable fact that the root system of $\mathscr{I}_4$, as it appears in Table 5.2, is *itself* a group (Exercise 5.16), a subgroup of the group of unit quaternions. Under other

Base	Group	$\|\Delta\|$	Root system Δ
A_n	$\mathscr{A}_n$	$n^2 + n$	$\pm(e_i - e_j)$, $1 \le j < i \le n + 1$.
B_n	$\mathscr{B}_n$	$2n^2$	$\pm e_i$, $1 \le i \le n$; $\pm e_i \pm e_j$, $1 \le j < i \le n$.
D_n	$\mathscr{D}_n$	$2n(n-1)$	$\pm e_i \pm e_j$, $1 \le j < i \le n$.
H_2^n	$\mathscr{H}_2^n$	$2n$	$(\cos j\pi/n, \sin j\pi/n)$, $0 \le j \le 2n - 1$.
G_2	$\mathscr{G}_2$	12	$\pm(e_i - e_j)$, $1 \le j < i \le 3$; $\pm(1, -2, 1)$, $\pm(-2, 1, 1)$, $\pm(1, 1, -2)$.
I_3	$\mathscr{I}_3$	30	$\pm e_i$, $1 \le i \le 3$; $\beta(\pm(2\alpha + 1), \pm 1, \pm 2\alpha)$, and all even permutations of coordinates.
I_4	$\mathscr{I}_4$	120	$\pm e_i$, $1 \le i \le 4$; $(1/2)(\pm 1, \pm 1, \pm 1, \pm 1)$; $\beta(\pm 2\alpha, 0, \pm(2\alpha + 1), \pm 1)$, and all even permutations of coordinates.
F_4	$\mathscr{F}_4$	48	$\pm e_i$, $1 \le i \le 4$; $\pm e_i \pm e_j$, $1 \le j < i \le 4$; $(1/2)\sum_1^4 \varepsilon_i e_i, \varepsilon_i = \pm 1$.
E_8	$\mathscr{E}_8$	240	$\pm e_i \pm e_j$, $1 \le j < i \le 8$; $(1/2)\sum_1^8 \varepsilon_i e_i$, $\varepsilon_i = \pm 1$, $\Pi_1^8 \varepsilon_i = -1$.
E_7	$\mathscr{E}_7$	126	Those roots of $\mathscr{E}_8$ orthogonal to $u = (1/2)(1, 1, 1, 1, 1, 1, 1, -1)$.
E_6	$\mathscr{E}_6$	72	Those roots of $\mathscr{E}_7$ orthogonal to $r_8 = e_8 - e_7$.

Table 5.2

circumstances that fact can be used in the construction of $\mathscr{I}_4$ (see [27], p. 308). A thorough discussion of finite groups of unit quaternions is given in [12].

We may now describe all finite subgroups of $\mathscr{O}(V)$ that are generated by reflections. We summarize in a theorem.

Theorem 5.3.1

If $\mathscr{G}$ is a finite subgroup of $\mathscr{O}(V)$ that is generated by reflections, then V may be written as the orthogonal direct sum of $\mathscr{G}$-invariant subspaces $V_0, V_1, \ldots, V_k$, with the following properties:

(a) If $\mathscr{G}_i = \{T|V_i : T \in \mathscr{G}\}$, then $\mathscr{G}_i \le \mathscr{O}(V_i)$ and $\mathscr{G}$ is isomorphic with $\mathscr{G}_0 \times \mathscr{G}_1 \times \cdots \times \mathscr{G}_k$.

(b) $\mathscr{G}_0$ consists only of the identity transformation on V_0.

(c) Each $\mathscr{G}_i$, $i \ge 1$, is one of the groups

$$\mathscr{A}_n, n \ge 1; \mathscr{B}_n, n \ge 2; \mathscr{D}_n, n \ge 4; \mathscr{H}_2^n, n \ge 5, n \ne 6;$$

$$\mathscr{G}_2; \mathscr{I}_3; \mathscr{I}_4; \mathscr{F}_4; \mathscr{E}_6; \mathscr{E}_7; \text{ or } \mathscr{E}_8.$$

The group $\mathscr{G}$ is a Coxeter group if and only if $V_0 = 0$, and $\mathscr{G}$ is crystallographic if and only if $\mathscr{H}_2^n$, $\mathscr{I}_3$, and $\mathscr{I}_4$ do not appear among the direct factors $\mathscr{G}_i$.

5.4 ORDERS OF IRREDUCIBLE COXETER GROUPS

The orders of the groups $\mathscr{A}_n$, $\mathscr{B}_n$, and $\mathscr{D}_n$ were computed in Section 5.3, and it was observed in Chapter 2 that the order of $\mathscr{H}_2^n$ is $2n$. Since $\mathscr{G}_2$ is just the dihedral group $\mathscr{H}_2^6$, it has order 12. The computation of the orders of the remaining irreducible Coxeter groups is less straightforward and requires some preparation.

We continue to assume that $\mathscr{G}$ is a Coxeter group (possibly reducible). As in Chapter 4 we shall denote by $\{s_1, \ldots, s_n\}$ the basis dual to $\Pi = \{r_1, \ldots, r_n\}$. The open half-space containing r_i determined by the hyperplane $\mathscr{P}_i = r_i^{\perp}$ will be denoted by L_i; i.e.,

$$L_i = \{x \in V : (x, r_i) > 0\}.$$

Theorem 5.4.1 (Witt, [27], p. 294)

If $\mathscr{H}$ is the subgroup of $\mathscr{G}$ leaving s_i fixed, then

$$\mathscr{H} = \langle S_1, \ldots, S_{i-1}, S_{i+1}, \ldots, S_n \rangle.$$

Proof

There is no loss of generality in assuming that $i = n$. Set $\mathscr{K} = \langle S_1, \ldots, S_{n-1} \rangle$. Then clearly $\mathscr{K} \leq \mathscr{H}$, since $s_n \perp r_i$, $1 \leq i \leq n-1$, and so $S_i s_n = s_n$. Let X_1 be the $(n-1)$ sphere of radius $\|s_n\|$ centered at the origin; i.e.,

$$X_1 = \{x \in V : \|x\| = \|s_n\|\}.$$

Let X_2 be an $(n-1)$ sphere of radius d centered at s_n, where d is sufficiently small so that $X_2 \subseteq L_n$ [$s_n \in L_n$ since $(s_n, r_n) = 1$, and L_n is an open set]. Then $X = X_1 \cap X_2$ is an $(n-2)$ sphere of radius $(d/2\|s_n\|)\sqrt{4\|s_n\|^2 - d^2}$ centered at the point $(1 - d^2/2\|s_n\|^2)s_n$ (see Exercise 5.13). Since $\mathscr{H} \leq \mathscr{O}(V)$ and $\mathscr{H}$ leaves s_n fixed, the sphere X is invariant under $\mathscr{H}$ (and also, of course, invariant under the subgroup $\mathscr{K}$).

Choose a point $x_0 \in L_1 \cap L_2 \cap \cdots \cap L_{n-1}$ that is not fixed by any nonidentity transformation in $\mathscr{H}$. Then the Fricke–Klein construction of Chapter 3 can be applied, using the point x_0, to construct fundamental regions $F(\mathscr{K})$ and $F(\mathscr{H})$ for $\mathscr{K}$ and $\mathscr{H}$ in V. Since $\mathscr{K} \leq \mathscr{H}$, we have $F(\mathscr{H}) \subseteq F(\mathscr{K})$. The half-spaces $L_1, \ldots, L_{n-1}$ are obtained when the reflections $S_1, \ldots, S_{n-1}$ are applied to x_0, so $F(\mathscr{K}) \subseteq L_1 \cap \cdots \cap L_{n-1}$. As indicated in Chapter 3, we obtain fundamental regions for $\mathscr{K}$ and $\mathscr{H}$ in X by setting $F_X(\mathscr{K}) = F(\mathscr{K}) \cap X$ and $F_X(\mathscr{H}) = F(\mathscr{H}) \cap X$. By Theorem 4.2.4 the set $F(\mathscr{G}) = L_1 \cap \cdots \cap L_n$ is a fundamental region for $\mathscr{G}$ in V. Since $X \subseteq L_n$, we have $F_X(\mathscr{K}) \subseteq L_1 \cap \cdots \cap L_{n-1} \cap X \subseteq F(\mathscr{G})$. If $F_X(\mathscr{H}) \neq F_X(\mathscr{K})$, then we may choose $x \in F_X(\mathscr{K}) \setminus F_X(\mathscr{H})$, $y \in F_X(\mathscr{H})$, and $T \in \mathscr{H}$, satisfying $Ty = x$. But that is impossible, since x and y are distinct points in the fundamental region $F(\mathscr{G})$. Thus $F_X(\mathscr{H}) = F_X(\mathscr{K})$, from which it follows that $|\mathscr{H}| = |\mathscr{K}|$, and hence that $\mathscr{H} = \mathscr{K}$, as desired.

Proposition 5.4.2

Suppose that $\mathscr{G}$ is irreducible and that its Coxeter graph G has no marks over its branches. Then $\mathscr{G}$ is transitive as a permutation group on its root system Δ.

Proof

Since $\Delta = \{Tr_i : T \in \mathscr{G}, r_i \in \Pi\}$, it will suffice to prove that if $r_i, r_j \in \Pi$, then $Tr_i = r_j$ for some $T \in \mathscr{G}$. Suppose that r_i and r_j correspond to nodes of G that are adjacent, i.e., joined by a branch. Then $(r_i, r_j)/\|r_i\| \, \|r_j\| = -1/2$, and

$$S_iS_jr_i = S_i(r_i + r_j) = -r_i + (r_j + r_i) = r_j.$$

Since roots corresponding to adjacent nodes are $\mathscr{G}$-equivalent, and since G is connected, by Proposition 5.1.4, it follows that any two roots in Π are $\mathscr{G}$-equivalent, and the proof is complete.

The proof of Proposition 5.4.2 will also show that $\mathscr{I}_3$ and $\mathscr{I}_4$ are transitive on their root systems if we can show that the roots $r_1, r_2 \in \Pi$, with $(r_1, r_2) = -\cos \Pi/5$, can be interchanged by elements of $\mathscr{I}_3$ and $\mathscr{I}_4$, respectively. It can be checked directly that $(S_1S_2)^2r_1 = r_2$. It is perhaps simpler and more illuminating to observe that $\langle S_1, S_2 \rangle$ is dihedral of order 10, and then to verify that $S_1S_2S_1S_2r_1 = r_2$ by following the arrows in Figure 5.6.

The group $\mathscr{F}_4$ is not transitive on its root system Δ. A modified version of the algorithm used above to construct all roots from the roots in a base can be used to determine the orbits of Δ. If we begin with a single root in the base of $\mathscr{F}_4$, say r_1, and apply to it every S_i for which $(r_1, r_i) \neq 0$, then apply every S_i to each resulting root r for which $(r, r_i) \neq 0$, and continue in this fashion, then the set of roots obtained is clearly $\mathrm{Orb}(r_1)$. A straightforward application of this procedure shows that Δ has two orbits: $\mathrm{Orb}(r_1)$ is the set of 24 roots of the form $\pm e_i \pm e_j$, and $\mathrm{Orb}(r_4)$ is the set of 24 roots of the forms $\pm e_i$, $(1/2)\Sigma \pm e_i$.

We may now compute the orders of the remaining groups. In each case we shall find a root $r \in \Delta$ that is orthogonal to all but one of the roots in the base Π given in Table 5.1. If $r_i \in \Pi$ is the fundamental root *not* orthogonal to r, then r is a scalar multiple of the dual basis vector s_i.

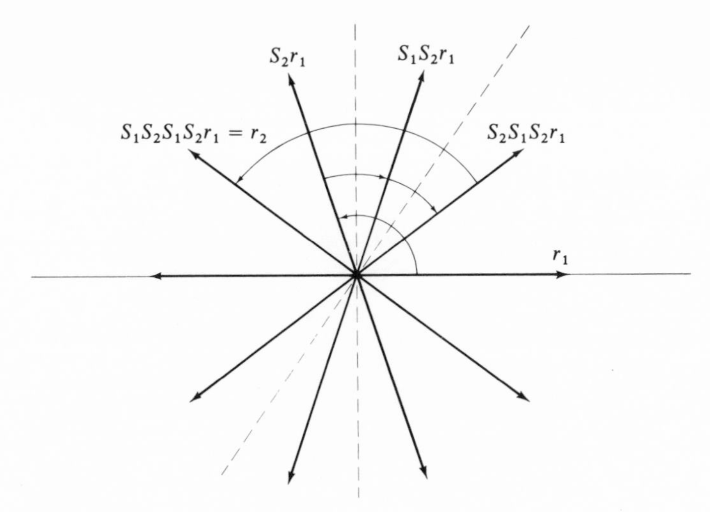

Figure 5.6

Thus the stabilizer Stab(r) is the subgroup of $\mathscr{G}$ generated by all fundamental reflections except S_i, by Witt's theorem. In each case |Stab(r)| and |Orb(r)| will be known, so an application of Proposition 1.2.1 completes the computation of $|\mathscr{G}|$.

Let us illustrate with the case of $\mathscr{F}_4$. The root $r = e_4 - e_3$ is orthogonal to r_1, r_2, and r_3 (see Table 5.1), so $r = \lambda s_4$ for some scalar λ, and $\text{Stab}(r) = \text{Stab}(s_4) = \langle S_1, S_2, S_3 \rangle$ by Theorem 5.4.1. Since the graph of $\{r_1, r_2, r_3\}$ is just B_3, we see that $\text{Stab}(r) = \mathscr{B}_3$, and so $|\text{Stab}(r)| = 2^3 \cdot 3! = 48$. As we observed above each orbit of Δ has 24 roots, so by Proposition 1.2.1 we have

$$\begin{aligned} |\mathscr{F}_4| &= [\mathscr{F}_4 : \text{Stab}(r)]|\text{Stab}(r)| \\ &= |\text{Orb}(r)|\,|\mathscr{B}_3| = 24 \cdot 48 = 1152. \end{aligned}$$

The procedure is the same for all the remaining cases. The remaining groups are transitive on their root systems, by Proposition 5.4.2 and the remarks following its proof, so Orb(r) is all of Δ. In Table 5.3 are listed the group $\mathscr{G}$, the root $r = \lambda s_i$, the stabilizer of r and its order, and the cardinality of Orb(r).

The order of each group in Table 5.3 is obtained by multiplying the numbers in the last two columns. In summary we tabulate all the irreducible Coxeter groups with their orders in Table 5.4.

$\mathscr{G}$	i	$r = \lambda s_i$	Stab(r)	|Stab(r)|	|Orb(r)|
$\mathscr{I}_3$	2	$\beta(1, 2\alpha, 2\alpha + 1)$	$\mathscr{A}_1 \times \mathscr{A}_1$	4	30
$\mathscr{I}_4$	4	e_4	$\mathscr{I}_3$	120	120
$\mathscr{F}_4$	4	$e_4 - e_3$	$\mathscr{B}_3$	48	24
$\mathscr{E}_6$	1	$e_7 + e_8$	$\mathscr{A}_5$	$6!$	72
$\mathscr{E}_7$	2	$e_1 + e_8$	$\mathscr{D}_6$	$2^5 \cdot 6!$	126
$\mathscr{E}_8$	8	$(1/2)(\Sigma_1^7 e_i - e_8)$	$\mathscr{E}_7$	$2^5 \cdot 6! \cdot 126$	240

Table 5.3

$\mathscr{G}$	$\|\mathscr{G}\|$	$\mathscr{G}$	$\|\mathscr{G}\|$
$\mathscr{A}_n$	$(n + 1)!$	$\mathscr{I}_3$	$2^3 \cdot 3 \cdot 5$
$\mathscr{B}_n$	$2^n \cdot n!$	$\mathscr{I}_4$	$2^6 \cdot 3^2 \cdot 5^2$
$\mathscr{D}_n$	$2^{n-1} \cdot n!$	$\mathscr{E}_6$	$2^7 \cdot 3^4 \cdot 5$
$\mathscr{H}_2^n$	$2n$	$\mathscr{E}_7$	$2^{10} \cdot 3^4 \cdot 5 \cdot 7$
$\mathscr{G}_2$	12	$\mathscr{E}_8$	$2^{14} \cdot 3^5 \cdot 5^2 \cdot 7$
$\mathscr{F}_4$	$2^7 \cdot 3^2$		

Table 5.4

Exercises

5.1 Suppose that $\mathscr{G}_1 \leq \mathscr{O}(V_1)$ and $\mathscr{G}_2 \leq \mathscr{O}(V_2)$ are Coxeter groups, and suppose that Π_1 is a t_1-base for $\mathscr{G}_1$ and Π_2 is a t_2-base for $\mathscr{G}_2$. Set $V = V_1 \oplus V_2$ and view $\mathscr{G}_1 \times \mathscr{G}_2$ as a group of transformations on V in the obvious fashion. Show that

$$\mathscr{G} = \mathscr{G}_1 \times \mathscr{G}_2 \leq \mathscr{O}(V),$$

and that $\mathscr{G}$ is a Coxeter group with $\Pi_1 \times \Pi_2$ as a t-base, where $t = (t_1, t_2) \in V$.

5.2 Suppose that $\{x_1, \ldots, x_n\}$ and $\{y_1, \ldots, y_n\}$ are bases of V and that $(x_i, x_j) = (y_i, y_j)$ for all i and j. Define a linear transformation $T: V \to V$ by setting $Tx_i = y_i$, $1 \leq i \leq n$. Show that $T \in \mathscr{O}(V)$.

5.3 Show that a reducible Coxeter group is crystallographic if and only if each of its irreducible direct factors is crystallographic.

5.4 If the set $\{x_1, \ldots, x_k\} \subseteq V$ is linearly independent, show that it lies on one side of some hyperplane; i.e., $(t, x_i) > 0$, all i, for some $t \in V$. (Try $t \perp (x_i - x_1)$, $2 \leq i \leq k$.)

5.5 Show that $\det U_{3'} = \det Z_4 = 0$.

5.6 Show that for each dimension n there do not exist geometrically different irreducible Coxeter groups of the same order.

5.7 Using Theorem 5.3.1 we may write down all finite reflection groups in two and three dimensions, and these groups must appear among the groups listed in Theorems 2.2.1 and 2.5.2. Verify the following identifications:

$$\mathscr{A}_1 \times 1 = \mathscr{H}_2^1; \quad \mathscr{A}_1 \times \mathscr{A}_1 = \mathscr{H}_2^2; \quad \mathscr{A}_2 = \mathscr{H}_2^3; \quad \mathscr{B}_2 = \mathscr{H}_2^4;$$
$$\mathscr{G}_2 = \mathscr{H}_2^6; \quad \mathscr{A}_1 \times 1_2 = \mathscr{C}_3^2]\mathscr{C}_3^1; \quad \mathscr{A}_1 \times \mathscr{A}_1 \times 1 = \mathscr{H}_3^2]\mathscr{C}_3^2;$$
$$\mathscr{A}_2 \times 1 = \mathscr{H}_2^3]\mathscr{C}_2^3; \quad \mathscr{B}_2 \times 1 = \mathscr{H}_3^4]\mathscr{C}_3^4; \quad \mathscr{G}_2 \times 1 = \mathscr{H}_3^6]\mathscr{C}_3^6;$$
$$\mathscr{A}_1 \times \mathscr{A}_1 \times \mathscr{A}_1 = (\mathscr{H}_3^2)^*; \quad \mathscr{A}_1 \times \mathscr{A}_2 = \mathscr{H}_3^6]\mathscr{H}_3^3;$$
$$\mathscr{A}_1 \times \mathscr{B}_2 = (\mathscr{H}_3^4)^*; \quad \mathscr{A}_1 \times \mathscr{G}_2 = (\mathscr{H}_3^6)^*; \quad \mathscr{A}_3 = \mathscr{W}]\mathscr{T};$$
$$\mathscr{B}_3 = \mathscr{W}^*; \quad \mathscr{I}_3 = \mathscr{I}^*; \quad \mathscr{A}_1 \times \mathscr{H}_2^n = \mathscr{H}_3^{2n}]\mathscr{H}_3^n$$

if n is odd and $n \geq 5$; and $\mathscr{A}_1 \times \mathscr{H}_2^n = (\mathscr{H}_3^n)^*$ if n is even and $n \geq 8$.

5.8 (a) If $\mathscr{G}$ is an irreducible crystallographic Coxeter group, show that the orbits in Δ are the subsets consisting of roots of the same length.
(b) What are the orbits in Δ if $\mathscr{G} = \mathscr{H}_2^n$?

5.9 Suppose that $\mathscr{G}$ is irreducible and that $R: V \to V$ is a linear transformation, with $TR = RT$ for all $T \in \mathscr{G}$.
(a) (*Schur's Lemma*) Show that either $R = 0$ or else R is invertible. (*Hint*: ker R is a $\mathscr{G}$-invariant subspace of V.)
(b) Show that each root r of $\mathscr{G}$ is an eigenvector of R, and in particular show that R has a real eigenvalue.
(c) If λ is a real eigenvalue of R, show that $R = \lambda 1$.
(d) Conclude that the center of $\mathscr{G}$ is either trivial or else consists of just ± 1.

5.10 Since there is an element $T \in \mathscr{B}_n$ such that $Te_i = -e_i$, $1 \leq i \leq n$, it is clear that the center of $\mathscr{B}_n$ has order 2 (see Exercise 5.12).
(a) Show that $\mathscr{A}_n$, $\mathscr{D}_{2n+1}$, and $\mathscr{H}_2^{2n+1}$ have trivial centers.
(b) Show that $\mathscr{D}_{2n}$ and $\mathscr{H}_2^{2n}$ have centers of order 2.

5.11 For each group $\mathscr{G}$ in Table 5.3, denote Stab(r) by $\mathscr{H}$. If W is the subspace of V spanned by $\Pi \setminus \{r_i\}$, then W is invariant under $\mathscr{H}$. Let $\mathscr{K}$ be the group of restrictions to W of transformations in $\mathscr{H}$.
(a) Suppose that $-1 \in \mathscr{K}$, say $-1 = T|W$, $T \in \mathscr{H}$. Use the fact that $\{r_1, \ldots, r_n, r\} \setminus \{r_i\}$ is a basis for V to show that $TS_r = -1 \in \mathscr{G}$.
(b) If $-1 \in \mathscr{G}$, then $-S_r \in \mathscr{G}$, and $-S_r \in \text{Stab}(r) = \mathscr{H}$. Show that $-S_r|W = -1 \in \mathscr{K}$.

(c) Conclude that $\mathscr{E}_6$ has trivial center, and that $\mathscr{G}_2, \mathscr{I}_3, \mathscr{I}_4, \mathscr{F}_4, \mathscr{E}_7$, and $\mathscr{E}_8$ have centers of order 2.

5.12 Construct the root systems of $\mathscr{I}_3, \mathscr{F}_4$, and $\mathscr{I}_4$.

5.13 Let Y be a sphere of radius α centered at $y \in V$ and Z a sphere of radius β centered at $z \in V$, and set $d = d(z, y)$. Show that the $(n-2)$ sphere $X = Y \cap Z$ has radius

$$\gamma = \frac{[2(\alpha^2\beta^2 + (\alpha^2 + \beta^2)d^2) - (\alpha^4 + \beta^4 + d^4)]^{1/2}}{2d^2}$$

and center

$$x = \left(\frac{d^2 - \alpha^2 + \beta^2}{2d^2}\right)y + \left(\frac{d^2 + \alpha^2 - \beta^2}{2d^2}\right)z,$$

assuming that $X \neq \varnothing$.

5.14 Identify $x = (\lambda_1, \lambda_2, \lambda_3, \lambda_4) \in \mathscr{R}^4$ with the real quaternion $x = \lambda_1 + \lambda_2 i + \lambda_3 j + \lambda_4 k \in Q$ (see [19], pp. 87 and 328–329).
(a) Show that the inner product in $\mathscr{R}^4$ can be expressed in terms of addition, multiplication, and the adjoint operation in Q by means of the formula

$$(x, y) = (1/2)(xy^* + yx^*).$$

(b) If $r \in \mathscr{R}^4$ is a unit vector, show that $S_r x = -rx^*r$ for all $x \in \mathscr{R}^4$.

5.15 Suppose that $\mathscr{K}$ is a finite subgroup of the multiplicative group of nonzero quaternions.
(a) Show that $\|x\| = 1$ for all $x \in \mathscr{K}$.
(b) If $x \in \mathscr{K}$, show that $x^* \in \mathscr{K}$.
(c) If $|\mathscr{K}|$ is even, show that $-1 \in \mathscr{K}$.
(d) For each $x \in \mathscr{K}$ let $S_x \in \mathscr{O}(\mathscr{R}^4)$ be the reflection along x. If $|\mathscr{K}|$ is even, show that $S_x\mathscr{K} = \mathscr{K}$ for each $x \in \mathscr{K}$ (see Exercise 5.14).

5.16 (From [27].) Let $\mathscr{K}$ be the group generated by the quaternions.

$$i, \qquad x = \alpha - (1/2)i - \beta j, \qquad \text{and} \qquad y = (1/2)(1 + i + j + k),$$

where $\alpha = \cos \pi/5$ and $\beta = \cos 2\pi/5$.
(a) Show that $i^2 = (ix)^3 = x^5 = -1$ and that $y = x^2 ix^{-1}$.
(b) Show that $|\mathscr{K}| = 120$.
(c) Conclude that $\mathscr{K}$ is the root system of $\mathscr{I}_4$.

chapter 6

GENERATORS AND RELATIONS FOR COXETER GROUPS

In this chapter we assume that the reader is familiar with the elementary theory of free groups (see [16], pp. 91–94, for example). Readers familiar with generators and relations may wish to proceed immediately to page 85.

If $\mathscr{G}$ is an arbitrary group generated by a subset $\mathscr{S}$, then there is a homomorphism φ from a free group $\mathscr{F}$ of rank $|\mathscr{S}|$ onto $\mathscr{G}$, and so $\mathscr{G}$ is isomorphic with the quotient group $\mathscr{F}/\mathscr{H}$, where $\mathscr{H}$ is the kernel of φ. In fact, $\mathscr{F}$ may be chosen to be the free group based on the set $\mathscr{S}$ itself. If an element T of $\mathscr{S}$ is denoted by $\hat{T}$ when it is considered as an element of the free group $\mathscr{F}$ rather than as an element of $\mathscr{G}$, then the homomorphism is the map that results from setting $\varphi(\hat{T}) = T$ for all $\hat{T} \in \mathscr{S}$.

If $T \in \mathscr{G}$, then T can be written as a product

$$T = T_1^{\varepsilon_1} T_2^{\varepsilon_2} \cdots T_k^{\varepsilon_k},$$

where each $T_i \in \mathscr{S}$ and each ε_i is either $+1$ or -1. When we wish to call attention to the particular product of generators we shall say that T is represented by the *word*

$$W = T_1^{\varepsilon_1} \cdots T_k^{\varepsilon_k}.$$

The corresponding element of the free group $\mathscr{F}$ is the word $\hat{W} = \hat{T}^{\varepsilon_1} \cdots \hat{T}_k^{\varepsilon_k}$ and two words $\hat{W}_1$ and $\hat{W}_2$ representing elements of $\mathscr{G}$ will be considered the same if and only if the corresponding $\hat{W}_1$ and $\hat{W}_2$ are equal in $\mathscr{F}$. To allay any possible confusion it should perhaps be emphasized that $W \to \hat{W}$ is *not* a homomorphism.

Suppose that $\{R_\alpha : \alpha \in A\}$, where A is some index set, is a collection of words in $\mathscr{G}$ with the property that each $\hat{R}_\alpha \in \mathscr{H} = \ker \varphi$, and $\mathscr{H}$ is the smallest normal subgroup of $\mathscr{F}$ containing all $\hat{R}_\alpha$. Then each element of $\mathscr{H}$ is a product of conjugates (in $\mathscr{F}$) of the words $\hat{R}_\alpha$ and their inverses. In this situation it is said that $\mathscr{G}$ has the *presentation*

$$\langle T \in \mathscr{S} | R_\alpha = 1, \alpha \in A \rangle,$$

or that $\mathscr{G}$ has *generators* $\mathscr{S}$ and *relations* R_α, $\alpha \in A$. For example, $\mathscr{H}_2^n$ has the presentation

$$\langle S_1, S_2 | S_1^2 = 1, S_2^2 = 1, (S_1 S_2)^n = 1 \rangle.$$

If certain relations $U_\beta = 1$, $\beta \in B$, where each U_β is a word in the generators $T \in \mathscr{S}$, hold in $\mathscr{G}$, then another relation $T_1 T_2 \cdots T_k = 1$ in $\mathscr{G}$ is called a *consequence* of the relations $U_\beta = 1$ if and only if the word $\hat{T}_1 \cdots \hat{T}_k$ is a product of conjugates of the words $\hat{U}_\beta$ and their inverses in the free group $\mathscr{F}$. Thus if *every* relation $T_1 \cdots T_k = 1$ in $\mathscr{G}$ is a consequence of the relations $U_\beta = 1$, it is clear that $\mathscr{G}$ has a presentation

$$\langle T \in \mathscr{S} | U_\beta = 1, \beta \in B \rangle.$$

We wish to show that if a relation $W = 1$ may be reduced to the relation $1 = 1$ by successive applications of the relations $U_\beta = 1$, then $W = 1$ is a consequence of the relations $U_\beta = 1$.

More explicitly, set $W_1 = W$ and suppose that the relation

$$W_1 = T_1 \cdots T_k V_1 T_{k+1} \cdots T_m = 1$$

is replaced by

$$W_2 = T_1 \cdots T_k V_2 T_{k+1} \cdots T_m = 1,$$

where V_1 and V_2 are words related by $V_1 = V_2 U_1$, $U_1 = 1$ being one of the relations $U_\beta = 1$. Then in the free group $\mathscr{F}$ we have

$$\begin{aligned}\hat{W}_1 &= \hat{T}_1 \cdots \hat{T}_k \hat{V}_1 \hat{T}_{k+1} \cdots \hat{T}_m = \hat{T}_1 \cdots \hat{T}_k \hat{V}_2 \hat{U}_1 \hat{T}_{k+1} \cdots \hat{T}_m \\ &= \hat{T}_1 \cdots \hat{T}_k \hat{V}_2 \hat{T}_{k+1} \cdots \hat{T}_m (\hat{T}_{k+1} \cdots \hat{T}_m)^{-1} \hat{U}_1 \hat{T}_{k+1} \cdots \hat{T}_m \\ &= (\hat{T}_1 \cdots \hat{T}_k \hat{V}_2 \hat{T}_{k+1} \cdots \hat{T}_m)(\hat{X}_1^{-1} \hat{U}_1 \hat{X}_1) = \hat{W}_2 (\hat{X}_1^{-1} \hat{U}_1 \hat{X}_1).\end{aligned}$$

Applying the same procedure to the relation $W_2 = 1$ we obtain

$$\hat{W}_2 = \hat{W}_3 (\hat{X}_2^{-1} \hat{U}_2 \hat{X}_2),$$

and hence

$$\hat{W}_1 = \hat{W}_3(\hat{X}_2^{-1}\hat{U}_2\hat{X}_2)(\hat{X}_1^{-1}\hat{U}_1\hat{X}_1).$$

Continuing, we obtain

$$\hat{W}_i = \hat{W}_{i+1}(\hat{X}_i^{-1}\hat{U}_i\hat{X}_i),$$

and so

$$\hat{W}_1 = \hat{W}_{i+1}(\hat{X}_i^{-1}\hat{U}_i\hat{X}_i)\cdots(\hat{X}_1^{-1}\hat{U}_1\hat{X}_1),$$

$i = 1, 2, 3, \ldots$. By hypothesis some W_{u+1} is just the identity element 1 of $\mathscr{G}$, so that W_{u+1} is the empty word in $\mathscr{F}$; so

$$\hat{W}_1 = (\hat{X}_u^{-1}\hat{U}_u\hat{X}_u)\cdots(\hat{X}_1^{-1}\hat{U}_1\hat{X}_1),$$

proving our contention.

Thus in order to show that $\mathscr{G}$ has a presentation $\langle T \in \mathscr{S} | U_\beta = 1, \beta \in B \rangle$, it is sufficient to show that every relation $W = 1$ in $\mathscr{G}$ may be reduced to the relation $1 = 1$ by successive applications of the relations $U_\beta = 1$.

Suppose now until further notice that $\mathscr{G}$ *is a Coxeter group in* $\mathcal{O}(V)$, *with base* $\Pi = \{r_1, \ldots, r_n\}$. Then $\mathscr{G}$ is generated by the set $\{S_1, \ldots, S_n\}$ of fundamental reflections. By Proposition 5.1.1 there are positive integers p_{ij} such that $(S_iS_j)^{p_{ij}} = 1$ for each pair i and j of subscripts. Our goal is to show that $\mathscr{G}$ has the presentation

$$\langle S_1, \ldots, S_n | \quad (S_iS_j)^{p_{ij}} = 1, 1 \leq i \leq j \leq n \rangle,$$

using the ideas above.

If $T \in \mathscr{G}$, then T may be represented (in several ways) as a product $S_{i_1} \cdots S_{i_k}$ of fundamental reflections. If $S_{i_1} \cdots S_{i_k}$ is a word representing T with the property that there is no word representing T having fewer than k fundamental reflections as factors, then we shall say that T has *length k*, and write $l(T) = k$. We agree that $l(1) = 0$.

Proposition 6.1.1 (Iwahori [12])

If $T \in \mathscr{G}$, then

$$l(TS_i) = \begin{cases} l(T) + 1 & \text{if } Tr_i \in \Delta^+, \\ l(T) - 1 & \text{if } Tr_i \in \Delta^-, \end{cases}$$

for each fundamental reflection S_i.

Proof

For each $R \in \mathscr{G}$ denote by $n(R)$ the number of positive roots that are sent to negative roots by R; i.e., $n(R) = |R(\Delta^+) \cap \Delta^-|$. By Proposition 4.1.9, $S_i(\Delta^+ \setminus \{r_i\}) = \Delta^+ \setminus \{r_i\}$, so

$$TS_i(\Delta^+ \setminus \{r_i\}) = T(\Delta^+ \setminus \{r_i\}).$$

Thus TS_i and T send the same number of positive roots different from r_i to negative roots. If $Tr_i \in \Delta^+$, then $TS_i r_i = -Tr_i \in \Delta^-$, so r_i is sent to a positive root by T but to a negative root by TS_i. As a result, $n(TS_i) = n(T) + 1$. Similarly, if $Tr_i \in \Delta^-$, then $TS_i r_i \in \Delta^+$ and $n(TS_i) = n(T) - 1$. To complete the proof let us show that $n(T) = l(T)$.

It is an immediate consequence of what has just been proved that $n(R) \leq l(R)$ for all $R \in \mathscr{G}$, for if $l(R) = k$, then R is a product of k fundamental reflections. Explicitly, if $R = S_{i_1} \cdots S_{i_k}$, then

$$\begin{aligned} n(S_{i_1}) &= 1, \\ n(S_{i_1}S_{i_2}) &= n(S_{i_1}) \pm 1 \leq 2, \\ n(S_{i_1}S_{i_2}S_{i_3}) &= n(S_{i_1}S_{i_2}) \pm 1 \leq 2 + 1 = 3, \\ &\vdots \\ n(S_{i_1} \cdots S_{i_k}) &= n(S_{i_1} \cdots S_{i_{k-1}}) \pm 1 \leq (k-1) + 1 = k. \end{aligned}$$

We may now show by induction on $n = n(T)$ that $l(T) = n(T)$. If $n = 0$ then $T = 1$, by Proposition 4.2.3, so $l(T) = 0$ as well. Suppose that $n \geq 1$ and that $n(R) = l(R)$ for all $R \in \mathscr{G}$ with $n(R) < n$. Choose $r \in \Delta^+$ for which $Tr \in \Delta^-$. Then $r = \Sigma_j \lambda_j r_j$, with all $\lambda_j \geq 0$, and $Tr = \Sigma_j \lambda_j Tr \in \Delta^-$, so some Tr_j must be a negative root. Thus $n(TS_j) = n(T) - 1 = n - 1$. By the induction hypothesis $l(TS_j) = n(TS_j) = n - 1$. But

$$l(T) = l((TS_j)S_j) \leq l(TS_j) + 1 = (n-1) + 1 = n.$$

Since the reverse inequality always holds, we have $l(T) = n = n(T)$, and the proposition is proved.

If S_i and S_j are fixed fundamental reflections and m is a nonnegative integer, let us denote by $(S_iS_j \cdots)_m$ the word $S_iS_jS_iS_j \cdots$ having m alternating factors S_i and S_j, beginning on the left with S_i. If $m = 0$ we agree, of course, that $(S_iS_j \cdots)_m = 1$. Similarly, $(\cdots S_iS_j)_m$ is the word with m alternating factors, with S_j at the extreme right. On occasion we shall also use the symbol $(\cdots S_iS_j \cdots)_m$ to denote a word with m alternating factors, where the context will determine which reflection occurs at either end of the word.

Proposition 6.1.2

If S_i and S_j are fundamental reflections in $\mathscr{G}$ and $1 \leq m \leq p_{ij}$, then

$$(\cdots S_i S_j)_{m-1} r_i \in \Delta^+.$$

Proof

If $i = j$, then $p_{ij} = m = 1$ and the result is trivial. Assume then that $i \neq j$. If the proposition is false, choose the smallest m for which $(\cdots S_i S_j)_{m-1} r_i \in \Delta^-$. Clearly $m > 1$. If m is even then

$$\begin{aligned}(\cdots S_i S_j)_{m-1} r_i &= (S_j \cdots S_i S_j)_{m-1} r_i \\ &= S_j(\cdots S_i S_j)_{m-2} r_i \in \Delta^-,\end{aligned}$$

and $(\cdots S_i S_j)_{m-2} r_i \in \Delta^+$ by the minimality of m. Thus by Proposition 4.1.9 we have $(\cdots S_i S_j)_{m-2} r_i = r_j$. By Proposition 4.1.1,

$$S_j = (\cdots S_i S_j)_{m-2} S_i (\cdots S_i S_j)_{m-2}^{-1},$$

so

$$S_j(\cdots S_i S_j)_{m-2} = (\cdots S_i S_j)_{m-2} S_i,$$

or

$$(S_j \cdots S_i S_j)_{m-1} = (S_i \cdots S_j S_i)_{m-1}.$$

But then $(S_i S_j \cdots)_{2m-2} = (S_i S_j)^{m-1} = 1$, contradicting the fact that $S_i S_j$ has order p_{ij}. The proof when m is odd is entirely analogous.

Proposition 6.1.3

Suppose that $T \in \mathscr{G}$, that i and j are fixed, and that $l(TS_i) = l(TS_j) = l(T) - 1$. Then $l(T(\cdots S_i S_j \cdots)_m) = l(T) - m$ if $0 \leq m \leq p_{ij}$.

Proof

We use induction on m. The result is trivial if $m = 0$ or $m = 1$. Suppose that $m \geq 2$ and that the conclusion holds if m is replaced by $m - 1$. By Proposition 6.1.1 we have $Tr_i, Tr_j \in \Delta^-$. By Proposition 6.1.2 the root $(\cdots S_i S_j)_{m-1} r_i$ is positive, so it is of the form $\alpha r_i + \beta r_j$ with α and β both nonnegative and not both zero. Thus

$$\begin{aligned}T(\cdots S_i S_j)_{m-1} r_i &= T(\alpha r_i + \beta r_j) \\ &= \alpha T r_i + \beta T r_j \in \Delta^-,\end{aligned}$$

and so

$$\begin{aligned}l(T(\cdots S_j S_i)_m) &= l(T(\cdots S_i S_j)_{m-1} S_i) \\ &= l(T(\cdots S_i S_j)_{m-1}) - 1 \\ &= l(T) - (m-1) - 1 = l(T) - m,\end{aligned}$$

by Proposition 6.1.1 and the induction hypothesis. Similarly, $l(T(\cdots S_iS_j)_m) = l(T) - m$.

If $W = S_{i_1} \cdots S_{i_k}$ is a word in $\mathcal{G}$, then each word $W_j = S_{i_1} \cdots S_{i_j}$, $0 \leq j \leq k$, will be called a *partial word* of W. If $j = 0$, then the word $W_j = W_0$ has no factors, and we agree as usual that $W_0 = 1$.

Theorem 6.1.4 (Coxeter [6])

Every relation $W = S_{i_1} \cdots S_{i_k} = 1$ in a Coxeter group $\mathcal{G}$ is a consequence of the relations $(S_iS_j)^{p_{ij}} = 1$, so $\mathcal{G}$ has the presentation

$$\langle S_1, \ldots, S_n | \quad (S_iS_j)^{p_{ij}} = 1, 1 \leq i \leq j \leq n \rangle.$$

Proof

Suppose that u is the maximal length of partial words of W. Then we may write W as $W_1S_iS_jW_2$, where $l(W_1S_i) = u$ and every partial word of W_1 has length less than u. Set $p = p_{ij}$, let $W' = W_1(S_jS_i \cdots)_{2p-2}W_2$, and observe that W and W' are equal, as elements of $\mathcal{G}$. With the exception of W_1S_i all partial words of W coincide, as group elements, with partial words of W'. In place of W_1S_i, W' has the partial words

$$W_1S_j, W_1S_jS_i, \ldots, W_1(S_jS_i \cdots)_{2p-3}.$$

Setting $T = W_1S_i$ and using the relation $S_i^2 = 1$, we see that the latter partial words of W' coincide, as group elements, with the words $T(S_iS_j \cdots)_m$, $2 \leq m \leq 2p - 2$. Since $l(T) = u$ is maximal, we have $l(TS_i) = l(TS_j) = l(T) - 1$. If $2 \leq m \leq p$, then $l(T(S_iS_j \cdots)_m) < u$ by Proposition 6.1.3. If $p < m \leq 2p - 2$, then $2 \leq 2p - m < p$, and

$$l(T(S_iS_j \cdots)_m) = l(T(S_jS_i \cdots)_{2p-m}) < u,$$

again by Proposition 6.1.3. Thus by applying the relation $(S_iS_j)^{p_{ij}} = 1$, we have replaced the original word W by another, W', all of whose partial words have length less than or equal to u, and having one fewer partial words of length u. Loosely speaking we have removed the first partial word of maximal length.

The procedure may now be repeated as many times as necessary until we arrive at the relation $1 = 1$, and the theorem is proved.

In order to illustrate the proof we introduce next a geometrical interpretation, due to Coxeter, of the words $S_{i_1} \cdots S_{i_k}$.

If F is the fundamental region described in Section 4.2, recall that $F^- \cap r_i^\perp$ is called the ith wall of F, and that S_i is the reflection through the ith wall. Each wall of the fundamental region TF, where $T \in \mathcal{G}$, is the image under T of a wall of F, and the image of the ith wall of F will be called the ith wall of TF.

Each word $S_{i_1}S_{i_2}\cdots S_{i_k}$ may be associated with a path in V in the following manner. First connect a point in F to a point in $S_{i_1}F$ by a path (e.g., a line segment) through the i_1st wall of F. Next connect that point to a point in an adjacent fundamental region by a path through the i_2nd wall of $S_{i_1}F$. Note that the i_2nd wall of $S_{i_1}F$ is the intersection of $S_{i_1}F^-$ with the hyperplane $S_{i_1}(r_{i_2}^{\perp})$, so the reflection that is actually applied at the second step to send $S_{i_1}F$ to the adjacent region is the reflection along $S_{i_1}r_{i_2}$, which, by Proposition 4.1.1, is $S_{i_1}S_{i_2}S_{i_1}^{-1}$. Thus the fundamental region reached at the second step is

$$(S_{i_1}S_{i_2}S_{i_1}^{-1})(S_{i_1}F) = S_{i_1}S_{i_2}F.$$

Next the point chosen in $S_{i_1}S_{i_2}F$ is connected to a point in an adjacent fundamental region by a path through the i_3rd wall of $S_{i_1}S_{i_2}F$. In this case the reflecting hyperplane is $S_{i_1}S_{i_2}(r_{i_3}^{\perp})$, so the resulting fundamental region is

$$(S_{i_1}S_{i_2})S_{i_3}(S_{i_1}S_{i_2})^{-1}(S_{i_1}S_{i_2}F) = S_{i_1}S_{i_2}S_{i_3}F.$$

Continuing in this fashion we obtain a path extending from F to $S_{i_1}\cdots S_{i_k}F$ corresponding to the word $S_{i_1}\cdots S_{i_k}$. In particular, if we have a relation $S_{i_1}\cdots S_{i_k} = 1$, then the corresponding path leads from F back to F, and we may choose the final end point to coincide with the initial point of the path. In other words we may take the path corresponding to $S_{i_1}\cdots S_{i_k}$ to be a *closed* path.

If X is a nonempty subset of V that is invariant under $\mathcal{G}$ and that is *pathwise connected*, then we may further restrict the path representing a word to be a path in X. In one of the illustrative examples below, for instance, $\mathcal{G}$ will be the group of symmetries of a tetrahedron and X will be the surface of the tetrahedron.

The lengths of elements T of $\mathcal{G}$ may be related to the fundamental regions of $\mathcal{G}$ as follows: Label the fundamental region F with 0, and for each $T \in \mathcal{G}$ label the fundamental region TF with the Roman numeral for $l(T)$. In practice the Roman numerals may be attached to the various fundamental regions as follows: F is labeled 0, then each region adjacent to F is labeled I, each region other than F that is adjacent to a region labeled I is labeled II, etc. In general, if a region has not yet been labeled and if it shares a wall with a labeled region, then it is labeled with the next higher Roman numeral.

We may now illustrate geometrically the proof of Coxeter's theorem by observing the effects that changes in the words have on paths representing them.

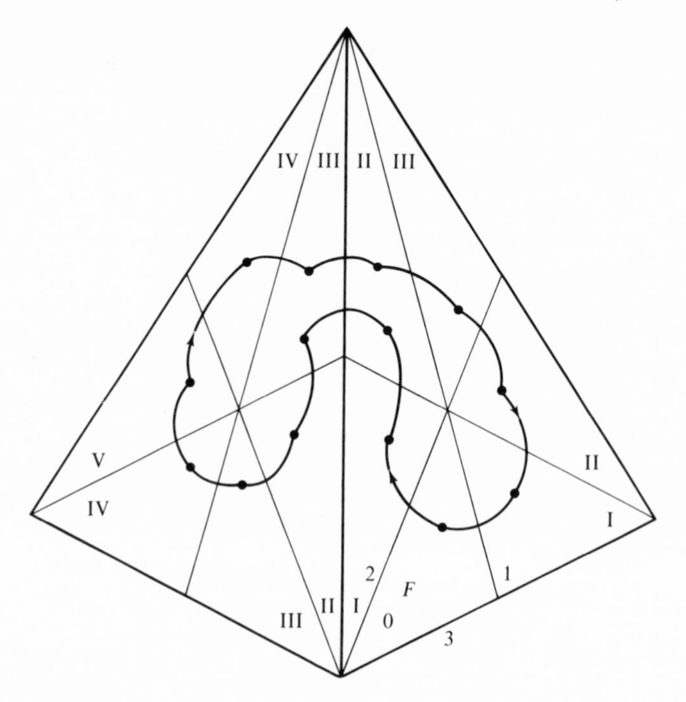

Figure 6.1

Suppose that $\mathscr{G} = \mathscr{A}_3 = \mathscr{W}\,]\,\mathscr{T}$, the group of all symmetries of a tetrahedron. Then

$$S_i^2 = (S_1S_2)^3 = (S_1S_3)^2 = (S_2S_3)^3 = 1.$$

The following relation holds in $\mathscr{G}$:

$$W = S_2S_1S_3S_1S_2S_1S_2S_1S_2S_3S_2S_1S_2S_1 = 1.$$

The corresponding closed path on the surface of the tetrahedron is shown in Figure 6.1. It is clear from the Roman numerals in the figure that 5 is the maximal length for partial words of W, and that there is just one partial word of length 5, viz. $S_2S_1S_3S_1S_2S_1S_2$. As in the proof of Theorem 6.1.4 we write $W = W_1S_2S_1W_2$ with $W_1 = S_2S_1S_3S_1S_2S_1$ and $W_2 = S_2S_3S_2S_1S_2S_1$. Applying the relation $(S_1S_2)^3 = 1$, we replace S_2S_1 by $S_1S_2S_1S_2$, thereby replacing $W = 1$ by the relation

$$\begin{aligned} W^{(1)} &= W_1S_1S_2S_1S_2W_2 \\ &= S_2S_1S_3S_1S_2S_1S_1S_2S_1S_2S_2S_3S_2S_1S_2S_1 \end{aligned}$$

The path corresponding to $W^{(1)}$ is shown in Figure 6.2(a).

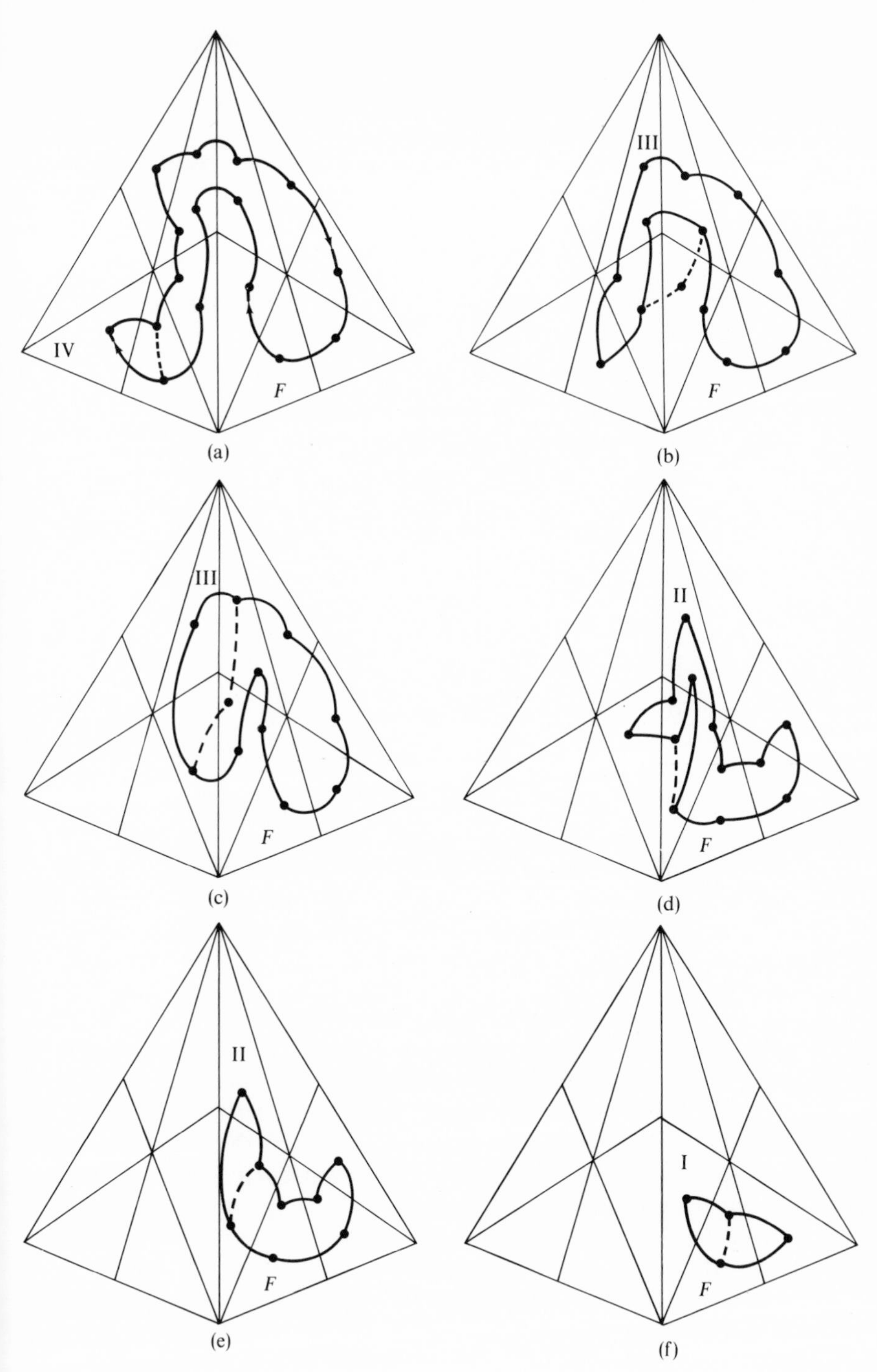

Figure 6.2

The word $W^{(1)}$ has no partial words of length 5 or greater, but it has two partial words of length 4—$S_2S_1S_3S_1S_2S_1$ and $S_2S_1S_3S_1S_2S_1S_1S_2S_1S_2$. Repeating the procedure we apply $S_1^2 = 1$ and obtain the relation

$$W^{(2)} = S_2S_1S_3S_1S_2S_2S_1S_2S_2S_3S_2S_1S_2S_1 = 1.$$

The change from the path of $W^{(1)}$ to the path of $W^{(2)}$ is indicated by the dashed line in Figure 6.2(a).

The next partial word of length 4 is removed by applying the relation $S_2^2 = 1$, and we obtain the relation

$$W^{(3)} = S_2S_1S_3S_1S_2S_2S_1S_3S_2S_1S_2S_1 = 1.$$

The corresponding path is shown (undashed) in Figure 6.2(b).

Continuing, we obtain the relations

$$W^{(4)} = S_2S_1S_1S_3S_2S_2S_1S_3S_2S_1S_2S_1 = 1,$$

$$W^{(5)} = S_2S_1S_1S_3S_1S_3S_2S_1S_2S_1 = 1,$$

$$W^{(6)} = S_2S_1S_1S_3S_3S_1S_2S_1S_2S_1 = 1,$$

$$W^{(7)} = S_2S_1S_1S_3S_3S_1S_1S_2S_1S_2S_2S_1 = 1,$$

$$W^{(8)} = S_2S_3S_3S_1S_1S_2S_1S_2S_2S_1 = 1,$$

$$W^{(9)} = S_2S_1S_1S_2S_1S_2S_2S_1 = 1,$$

$$W^{(10)} = S_2S_2S_1S_2S_2S_1 = 1,$$

$$W^{(11)} = S_2S_2S_1S_1 = 1, \qquad W^{(12)} = S_1S_1 = 1,$$

and finally $1 = 1$. The corresponding paths are shown in Figure 6.2.

In the example, each replacement of one word by another by means of applying a relation $(S_iS_j)^{p_{ij}} = 1$, with $i \neq j$, corresponds to replacing a path by another path that passes on the opposite side of a particular edge of a fundamental region, i.e., to "pulling" a path past an edge of the region. Likewise, an application of a relation $S_i^2 = 1$ corresponds to pulling a path through a wall of a fundamental region.

It is interesting to observe that if $p_{ij} > 2$, then the word W' actually has more fundamental reflections as factors than the word W it replaces. Geometrically, the path of W' is longer, in the sense that it reaches more fundamental regions, than the path of W. However, the path of W' is shorter than that of W in the more important sense that it does not wander so far from the initial fundamental region F.

For the further discussion of examples, the procedure in the proof of Theorem 6.1.4 may be modified so as to make it more efficient. As before, write $W = W_1S_iS_jW_2$, where W_1S_i is the first partial word of maximal length u. Then write W_1S_i as $W_3(\cdots S_jS_i)_v$, where the last factor of W_3 is neither S_i nor S_j (observe that $v < 2p_{ij}$). Next write S_jW_2 as $(S_jS_i\cdots)_wW_4$, where $v + w = 2p_{ij}$, if possible, or otherwise $v + w < 2p_{ij}$ and the first factor of W_4 is neither S_i nor S_j. Let W' be the word obtained from W by applying $(S_iS_j)^{p_{ij}} = 1$ and replacing $(\cdots S_jS_i)_v(S_jS_i\cdots)_w$ by $(\cdots S_iS_j\cdots)_k$, where $k = 2p_{ij} - v - w$. Then W' has one fewer partial words of length u than W (see Exercise 6.3).

To illustrate, suppose that $\mathscr{G} = \mathscr{B}_3 = \mathscr{W}^*$, so that $S_i^2 = (S_1S_2)^4 = (S_1S_3)^2 = (S_2S_3)^3 = 1$. The following relation holds in $\mathscr{G}$:

$$W = S_2S_3S_1S_2S_1S_3S_1S_2S_3S_2S_1S_3S_2S_1S_2S_1S_2S_1S_2S_3S_1S_2 = 1.$$

The corresponding path on the surface of the cube is shown in Figure 6.3(a). The maximal length of partial words of W is 6, and there are two

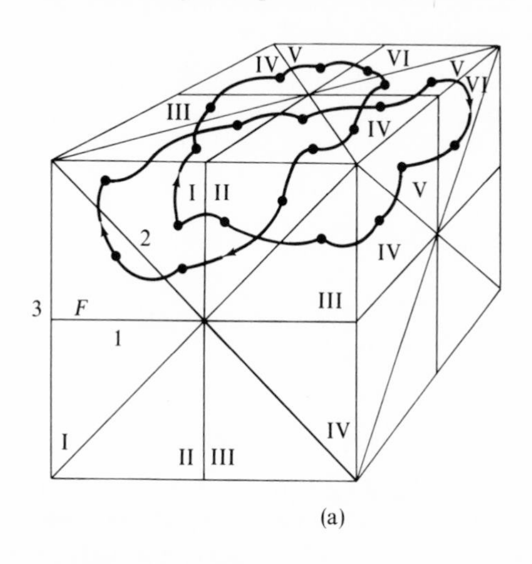

(a)

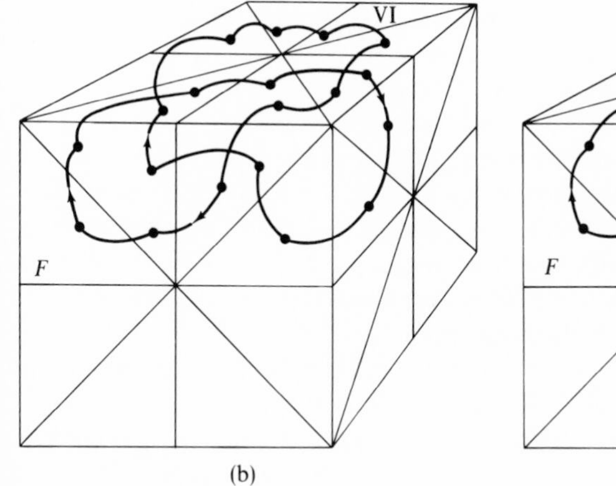

(b)

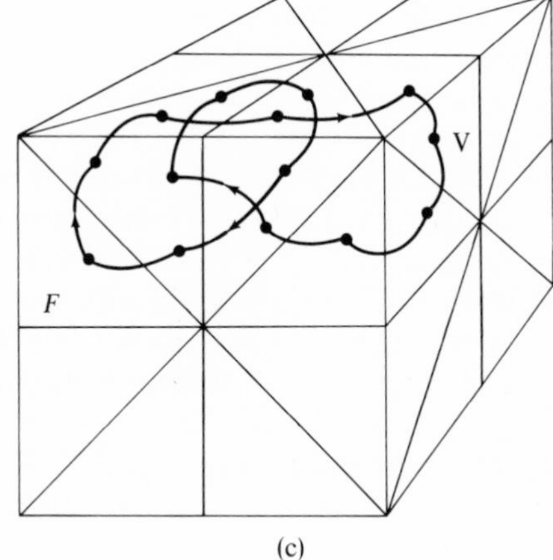

(c)

Figure 6.3

occurrences—$S_2S_3S_1S_2S_1S_3$ and $S_2S_3S_1S_2S_1S_3S_1S_2S_3S_2S_1S_3S_2S_1S_2S_1$. We write $W = W_3(S_1S_3)S_1W_4$, as indicated above, and apply $(S_1S_3)^2 = 1$, replacing $S_1S_3S_1$ by S_3, thereby replacing $W = 1$ by the relation

$$W^{(1)} = W_3S_3W_4 = 1.$$

The corresponding path is shown in Figure 6.3(b).

The word $W^{(1)}$ has $S_2S_3S_1S_2S_3S_2S_3S_2S_1S_3S_2S_1S_2S_1$ as its only partial word of length 6. We next apply $(S_1S_2)^4 = 1$, replacing $S_2S_1S_2S_1S_2S_1S_2$ by S_1, and obtain the relation

$$W^{(2)} = S_2S_3S_1S_2S_3S_2S_3S_2S_1S_3S_1S_3S_1S_2 = 1.$$

The path is shown in Figure 6.3(c).

The succeeding applications of the procedure and the drawing of the resulting paths are left as an exercise.

For the remainder of the chapter we shall suppose that $\mathscr{G}$ is a finite group having the presentation

$$\langle T_1, \ldots, T_n | (T_iT_j)^{p_{ij}} = 1, 1 \le i, j \le n \rangle,$$

where $p_{ii} = 1$ and $p_{ij} = p_{ji} \ge 2$ if $i \ne j$. We wish to establish the converse to Theorem 6.1.4, i.e., to show that $\mathscr{G}$ is (isomorphic with) a Coxeter group.

Set $\mathscr{S} = \{T_1, \ldots, T_n\}$, the set of generators of $\mathscr{G}$. If $\mathscr{S}$ is a disjoint union of nonempty sets $\mathscr{S}_1$ and $\mathscr{S}_2$, with $p_{ij} = 2$ whenever $T_i \in \mathscr{S}_1$ and $T_j \in \mathscr{S}_2$, then $\mathscr{G}$ is called *decomposable*. Otherwise $\mathscr{G}$ is called *indecomposable*.

Suppose that $\mathscr{G}$ is decomposable, with $\mathscr{S} = \mathscr{S}_1 \cup \mathscr{S}_2$. By relabeling, if necessary, we may assume that $\mathscr{S}_1 = \{T_1, \ldots, T_k\}$ and $\mathscr{S}_2 = \{T_{k+1}, \ldots, T_n\}$. Set $\mathscr{G}_i = \langle \mathscr{S}_i \rangle \le \mathscr{G}, i = 1, 2$. The (external) direct product $\mathscr{G}_1 \times \mathscr{G}_2$ has generators T'_i, where

$$T'_i = \begin{cases} (T_i, 1) & \text{if } 1 \le i \le k, \\ (1, T_i) & \text{if } k + 1 \le i \le n, \end{cases}$$

that satisfy the relations $(T'_iT'_j)^{p_{ij}} = 1$. Thus $\mathscr{G}_1 \times \mathscr{G}_2$ is a homomorphic image of $\mathscr{G}$, and $|\mathscr{G}_1 \times \mathscr{G}_2| \le |\mathscr{G}|$.

If $k + 1 \le i \le n$, then $T_i^{-1}\mathscr{G}_1T_i = \mathscr{G}_1$, since T_i commutes with each generator of $\mathscr{G}_1$. Thus $\mathscr{G}_1$ is a normal subgroup of $\mathscr{G}$ and, similarly, $\mathscr{G}_2$ is a normal subgroup of $\mathscr{G}$. Furthermore, $\mathscr{G}_1\mathscr{G}_2 = \mathscr{G}$, again since the respective generators commute with one another. Thus

$$\begin{aligned} |\mathscr{G}| = |\mathscr{G}_1\mathscr{G}_2| &= |\mathscr{G}_1|\,|\mathscr{G}_2|/|\mathscr{G}_1 \cap \mathscr{G}_2| \\ &= |\mathscr{G}_1 \times \mathscr{G}_2|/|\mathscr{G}_1 \cap \mathscr{G}_2| \le |\mathscr{G}|/|\mathscr{G}_1 \cap \mathscr{G}_2| \le |\mathscr{G}|. \end{aligned}$$

As a result, $\mathscr{G}_1 \cap \mathscr{G}_2 = 1$, and $\mathscr{G}$ is in fact the (internal) direct product of its subgroups $\mathscr{G}_1$ and $\mathscr{G}_2$.

If $\mathscr{K}_1$ and $\mathscr{K}_2$ are groups having presentations

$$\langle \mathscr{S}_1 | (T_i T_j)^{p_{ij}} = 1, 1 \leq i, j \leq k \rangle,$$
$$\langle \mathscr{S}_2 | (T_i T_j)^{p_{ij}} = 1, k + 1 \leq i \leq n \rangle,$$

respectively, then $\mathscr{G}_1$ and $\mathscr{G}_2$ are homomorphic images of $\mathscr{K}_1$ and $\mathscr{K}_2$; so $|\mathscr{G}_1| \leq |\mathscr{K}_1|$, $|\mathscr{G}_2| \leq |\mathscr{K}_2|$. But the presentation given for $\mathscr{G}$ is a presentation for $\mathscr{K}_1 \times \mathscr{K}_2$ (see Exercise 6.2), so

$$\mathscr{K}_1 \times \mathscr{K}_2 \cong \mathscr{G} = \mathscr{G}_1 \times \mathscr{G}_2.$$

It follows that $|\mathscr{G}_i| = |\mathscr{K}_i|$, and so $\mathscr{G}_i \cong \mathscr{K}_i$, $i = 1, 2$.

Summing up, we have shown that if $\mathscr{G}$ is decomposable, then it is a direct product of subgroups, each having a presentation of the same type that $\mathscr{G}$ has. If either of the direct factors is decomposable, it is also a direct product. Ultimately, we see that $\mathscr{G}$ is a direct product of indecomposable groups, each having a presentation of the same type that $\mathscr{G}$ has. If each of the indecomposable direct factors $\mathscr{G}_i$ of $\mathscr{G}$ is isomorphic with a Coxeter group $\mathscr{H}_i$, then $\mathscr{G}$ is isomorphic with the Coxeter group that is the direct product of the Coxeter groups $\mathscr{H}_i$. Consequently, it will be sufficient to show that indecomposable groups are isomorphic with Coxeter groups.

We assume for the remainder of the discussion that $\mathscr{G}$ is indecomposable.

If $\{e_1, \ldots, e_n\}$ is the usual basis for $\mathscr{R}^n$, define transformations $S_1, \ldots, S_n$ on $\mathscr{R}^n$ by setting

$$S_i e_j = e_j + 2(\cos \pi/p_{ij})e_i.$$

Set $\alpha_{ij} = -\cos \pi/p_{ij}$ and let A denote the $n \times n$ matrix (α_{ij}). Suppose that $x = \Sigma_j \lambda_j e_j \in \mathscr{R}^n$. If we set $a_i = (\alpha_{i1}, \alpha_{i2}, \ldots, \alpha_{in}) \in \mathscr{R}^n$, the ith row (or column) of A, then we have

$$S_i x = \Sigma_j \lambda_j S_i e_j = \Sigma_j \lambda_j (e_j - 2\alpha_{ij} e_i)$$
$$= \Sigma_j \lambda_j e_j - 2\,\Sigma_j \lambda_j \alpha_{ij} e_i = x - 2(x, a_i)e_i.$$

Set $\mathscr{P}_i = a_i^{\perp} \subseteq \mathscr{R}^n$; i.e.,

$$\mathscr{P}_i = \{x \in \mathscr{R}^n : (x, a_i) = 0\}.$$

If $x \in \mathscr{P}_i$, then $S_i x = x - 2(x, a_i)e_i = x$, and $S_i e_i = e_i - 2(e_i, a_i)e_i = e_i - 2\alpha_{ii}e_i = -e_i$. Since S_i leaves the elements of the hyperplane $\mathscr{P}_i$ pointwise fixed and carries e_i to its negative, we see that geometrically S_i is the reflection (not necessarily orthogonal) through $\mathscr{P}_i$ in the direction of e_i. Our procedure will be to modify the inner product on $\mathscr{R}^n$ so that $S_1, \ldots, S_n$ will become orthogonal transformations and generate a Coxeter group isomorphic with $\mathscr{G}$.

Denote by $\mathscr{H}$ the group of nonsingular transformations of $\mathscr{R}^n$ generated by $\{S_1, \ldots, S_n\}$.

Proposition 6.1.5

There is a homomorphism ψ from $\mathscr{G}$ onto $\mathscr{H}$ with $\psi(T_i) = S_i$, $1 \leq i \leq n$. In particular, $\mathscr{H}$ is a finite group.

Proof

We need only show that the generators $S_1, \ldots, S_n$ of $\mathscr{H}$ satisfy the relations in the presentation of $\mathscr{G}$ (see Exercise 6.1). Since $S_i e_i = -e_i$, we have for all $x \in \mathscr{R}^n$ that

$$\begin{aligned} S_i^2 x &= S_i(x - 2(x, a_i)e_i) = S_i x - 2(x, a_i)S_i e_i \\ &= x - 2(x, a_i)e_i + 2(x, a_i)e_i = x; \end{aligned}$$

so $S_i^2 = 1$.

In order to show that $(S_i S_j)^{p_{ij}} = 1$ when $i \neq j$, we lose no generality by taking $i = 1$ and $j = 2$. Since $\alpha_{11} = \alpha_{22} = 1$ and $0 \leq \alpha_{ij} < 1$ if $i \neq j$, it is immediate that a_1 and a_2 are linearly independent in $\mathscr{R}^n$. Thus their orthogonal complements $\mathscr{P}_1$ and $\mathscr{P}_2$ are different from one another, and $\mathscr{P}_1 \cap \mathscr{P}_2$ has dimension $n - 2$. Choose $x_1, \ldots, x_n \in \mathscr{R}^n$ such that $\{x_1, x_3, x_4, \ldots, x_n\}$ is a basis for $\mathscr{P}_1$ and $\{x_2, x_3, \ldots, x_n\}$ is a basis for $\mathscr{P}_2$ (and hence $\{x_3, \ldots, x_n\}$ is a basis for $\mathscr{P}_1 \cap \mathscr{P}_2$). Let us show that $\{e_1, e_2, x_3, \ldots, x_n\}$ is a basis for $\mathscr{R}^n$. If $\lambda_1 e_1 + \lambda_2 e_2 + \Sigma_{i=3}^n \lambda_i x_i = 0$ were a nontrivial dependence relation, then $\lambda_1 \neq 0$ and $\lambda_2 \neq 0$ since e_1, $e_2 \notin \mathscr{P}_1 \cap \mathscr{P}_2$. Thus $x = \lambda_1 e_1 + \lambda_2 e_2 \in \mathscr{P}_1 \cap \mathscr{P}_2$. But then $(x, a_1) = (x, a_2) = 0$, or

$$\begin{aligned} \lambda_1 + \alpha_{12}\lambda_2 &= 0, \\ \alpha_{12}\lambda_1 + \lambda_2 &= 0; \end{aligned}$$

so $\lambda_1 - \alpha_{12}^2\lambda_1 = 0$, or $\alpha_{12}^2 = 1 = \cos^2 \pi/p_{12}$, contradicting the fact that $p_{12} \geq 2$.

If $p_{12} = 2$, then $\alpha_{12} = -\cos \pi/2 = 0$, and so $S_1 e_2 = e_2$, $S_2 e_1 = e_1$. Thus

$$\begin{aligned} S_1 S_2 x &= x \qquad \text{if } x \in \mathscr{P}_1 \cap \mathscr{P}_2, \\ S_1 S_2 e_1 &= S_1 e_1 = -e_1, \end{aligned}$$

and

$$S_1 S_2 e_2 = S_1(-e_2) = -e_2.$$

It follows that $(S_1 S_2)^2 = (S_1 S_2)^{p_{12}} = 1$.

Finally, suppose that $p_{12} > 2$. With respect to the basis $\{e_1, e_2, x_3, \ldots, x_n\}$ the transformation $S_1 S_2$ is represented by the matrix

$$\left[\begin{array}{cc|c} 4\alpha_{12}^2 - 1 & -2\alpha_{12} & 0 \\ 2\alpha_{12} & -1 & \\ \hline \multicolumn{2}{c|}{0} & I_{n-2} \end{array}\right].$$

Easy computations show that the 2×2 matrix

$$B = \begin{bmatrix} 4\alpha_{12}^2 - 1 & -2\alpha_{12} \\ 2\alpha_{12} & -1 \end{bmatrix}$$

has eigenvalues $e^{2\pi i/p_{12}}$ and $e^{-2\pi i/p_{12}}$. Since $p_{12} > 2$, the eigenvalues of B are distinct and B is similar (over the complex field) to the matrix

$$C = \begin{bmatrix} e^{2\pi i/p_{12}} & 0 \\ 0 & e^{-2\pi i/p_{12}} \end{bmatrix}.$$

Since $C^{p_{12}} = I$, we conclude that $B^{p_{12}} = I$, and hence that $(S_1 S_2)^{p_{12}} = 1$, proving the proposition.

For each pair x and y of vectors in $\mathscr{R}^n$ set

$$C(x, y) = \Sigma\{(Tx, Ty) : T \in \mathscr{H}\}.$$

It is immediate that C is a symmetric bilinear form on $\mathscr{R}^n$. The associated quadratic form Q is positive definite since

$$Q(x) = C(x, x) = \Sigma \|Tx\|^2 > 0 \qquad \text{if } x \neq 0,$$

so C is an inner product on $\mathscr{R}^n$. Furthermore, if $R \in \mathscr{H}$ then

$$\begin{aligned} C(Rx, Ry) &= \Sigma\{(TRx, TRy) : T \in \mathscr{H}\} \\ &= \Sigma\{(Ux, Uy) : U \in \mathscr{H}\} = C(x, y), \end{aligned}$$

and C is invariant under all the transformations in $\mathscr{H}$. Equivalently, each $R \in \mathscr{H}$ is an orthogonal transformation with respect to the inner product C.

Proposition 6.1.6

Suppose that W is a subspace of $\mathscr{R}^n$ such that $TW \subseteq W$ for all $T \in \mathscr{H}$. Then either $W = \mathscr{R}^n$ or $W = 0$.

Proof

Suppose that $W \neq \mathscr{R}^n$, $W \neq 0$, and set

$$W' = \{x \in \mathscr{R}^n : C(x, y) = 0 \text{ for all } y \in W\}.$$

Then W' is the orthogonal complement of W with respect to the inner product C, so $\mathscr{R}^n = W \oplus W'$. If $T \in \mathscr{H}$ and $x \in W'$, then

$$C(Tx, y) = C(T^{-1}Tx, T^{-1}y) = C(x, T^{-1}y) = 0$$

if $y \in W$, since then also $T^{-1}y \in W$. Thus $TW' \subseteq W'$ for all $T \in \mathscr{H}$. If $e_i \in W$ and $e_j \notin W$, then $S_j e_i = e_i - 2\alpha_{ij} e_j \in W$; so $\alpha_{ij} = 0$, or $p_{ij} = 2$, for otherwise $e_j \in W$. The same reasoning shows that if $e_i \in W'$ and $e_j \notin W'$, then $p_{ij} = 2$. Since $\mathscr{G}$ is indecomposable, it follows that no e_i is in either W or W'. Write $e_1 = x + y$, with $x \in W$, $y \in W'$. Then $S_1 x = x - 2(x, a_1)e_1 \in W$, and

$$2(x, a_1)e_1 = x - S_1 x \in W;$$

so $(x, a_1) = 0$ and $S_1 x = x$, for otherwise $e_1 \in W$. Similarly, $S_1 y = y$; so

$$-e_1 = S_1 e_1 = S_1(x + y) = x + y = e_1,$$

a contradiction.

We define another bilinear form B on $\mathscr{R}^n$ by setting $B(x, y) = (Ax, y)$. Since A is a symmetric matrix, B is a symmetric form. Observe that

$$B(x, e_i) = (Ax, e_i) = (x, Ae_i) = (x, a_i)$$

for each i, so

$$S_i x = x - 2B(x, e_i)e_i$$

for all $x \in \mathscr{R}^n$. Also $B(e_i, e_i) = \alpha_{ii} = 1$.

In order to check that the form B is invariant under the transformations in $\mathscr{H}$, it will suffice to check that $B(S_i x, S_i y) = B(x, y)$ for each S_i, since the transformations S_i generate $\mathscr{H}$. Given $x, y \in \mathscr{R}^n$, we have

$$\begin{aligned} B(S_i x, S_i y) &= B(x - 2B(x, e_i)e_i, y - 2B(y, e_i)e_i) \\ &= B(x, y) - 2B(x, e_i)B(e_i, y) \\ &\quad - 2B(y, e_i)B(e_i, x) + 4B(x, e_i)B(y, e_i)B(e_i, e_i) \\ &= B(x, y), \end{aligned}$$

as desired.

Proposition 6.1.7 (Schur's Lemma)

Suppose that $\mathscr{H}$ is a group (or in fact any set) of linear transformations on a finite-dimensional vector space V over a field $\mathscr{F}$, and suppose that the only $\mathscr{H}$-invariant subspaces of V are 0 and V. If S is a nonzero linear transformation on V such that $ST = TS$ for all $T \in \mathscr{H}$, then S is nonsingular.

Proof

Let W be the null space of S. Then $W \neq V$ since $S \neq 0$. If $x \in W$ and $T \in \mathscr{H}$, then $STx = TSx = T0 = 0$, so $Tx \in W$. Thus W is $\mathscr{H}$-invariant, so $W = 0$, and S is nonsingular.

Corollary

If S has an eigenvalue λ in $\mathscr{F}$, then $S = \lambda 1$.

Proof

The transformation $S - \lambda 1$ also commutes with all $T \in \mathscr{H}$. If $S - \lambda 1$ were nonzero, it would be nonsingular by Schur's Lemma, contradicting the definition of an eigenvalue.

Suppose that $\mathscr{H} = \{R_1, R_2, \ldots, R_m\}$, and let R_i be represented by the matrix M_i with respect to the basis $\{e_1, \ldots, e_n\}$. If we set $P = \Sigma_{i=1}^m M_i^t M_i$, then P is easily seen to be symmetric and positive definite. In fact,

$$\begin{aligned} C(x, y) &= \Sigma_i(R_i x, R_i y) = \Sigma_i(M_i x, M_i y) \\ &= \Sigma_i(M_i^t M_i x, y) = (Px, y). \end{aligned}$$

Proposition 6.1.8

If $T \in \mathscr{H}$ is represented by the matrix M with respect to the basis $\{e_1, \ldots, e_n\}$, then

$$M(P^{-1}A) = (P^{-1}A)M.$$

Proof

Since the forms B and C are $\mathscr{H}$-invariant, we have

$$\begin{aligned} C(P^{-1}Ax, y) &= (PP^{-1}Ax, y) = (Ax, y) = B(x, y) \\ &= B(T^{-1}x, T^{-1}y) = B(M^{-1}x, M^{-1}y) \\ &= (AM^{-1}x, M^{-1}y) = (PP^{-1}AM^{-1}x, M^{-1}y) \\ &= C(P^{-1}AM^{-1}x, M^{-1}y) = C(MP^{-1}AM^{-1}x, y) \end{aligned}$$

for all $x,y \in \mathscr{R}^n$. Since C is an inner product, we may conclude that $P^{-1}A = MP^{-1}AM^{-1}$, and hence that $M(P^{-1}A) = (P^{-1}A)M$.

Theorem 6.1.9

The bilinear form B is a positive scalar multiple of the inner product C, so B is also an inner product on $\mathscr{R}^n$.

Proof

Since the matrix P^{-1} is positive definite, we may write $P^{-1} = NN^t$, where N is a nonsingular real matrix (see [1], p. 256). If $T \in \mathscr{H}$ is represented by the matrix M with respect to the basis $\{e_1, \ldots, e_n\}$, then T is represented by the matrix $N^{-1}MN$ with respect to the basis of columns of N. By Proposition 6.1.8 we have

$$(N^{-1}MN)(N^tAN) = N^{-1}M(P^{-1}A)N = N^{-1}(P^{-1}A)MN$$
$$= N^{-1}NN^tAMN = (N^tAN)(N^{-1}MN).$$

Thus the transformation S represented by the matrix N^tAN with respect to the basis of columns of N commutes with all $T \in \mathscr{H}$. Since N^tAN is symmetric, its eigenvalues are real; so by Proposition 6.1.6 and the Corollary to Schur's Lemma we see that $N^tAN = \lambda I$ for some nonzero scalar λ. Thus

$$A = \lambda(N^t)^{-1}N^{-1} = \lambda(NN^t)^{-1} = \lambda P,$$

so

$$B(x, y) = (Ax, y) = \lambda(Px, y) = \lambda C(x, y).$$

Since $1 = B(e_1, e_1) = \lambda C(e_1, e_1)$ and $C(e_1, e_1) > 0$, we see that $\lambda > 0$, and the theorem is proved.

Let us denote by V the vector space $\mathscr{R}^n$ endowed with the inner product B.

Theorem 6.1.10 (Coxeter)

The group $\mathscr{H}$ is a Coxeter subgroup of $\mathscr{O}(V)$, and $\mathscr{G}$ is isomorphic with $\mathscr{H}$.

Proof

The transformations S_i are orthogonal reflections of V since $S_ix = x - 2B(x, e_i)e_i$ for all x, and $r_i = e_i$ is a root of S_i. Since $\{r_1, \ldots, r_n\}$ is a basis for V, $\mathscr{H}$ is effective and hence is a Coxeter group. The quadratic form of the set of roots $\{r_1, \ldots, r_n\}$ has matrix A since

$$B(r_i, r_j) = \alpha_{ij} = -\cos \pi/p_{ij},$$

so $\{r_1, \ldots, r_n\}$ has a positive definite Coxeter graph. The graph is connected since $\mathscr{G}$ is indecomposable, so it must be one of the graphs in Figure 5.3. The construction of Section 5.3 may be applied to the roots $\{r_1, \ldots, r_n\}$, and we see that $\{S_1, \ldots, S_n\}$ are fundamental reflections for $\mathscr{H}$. But then by Theorem 6.1.4 $\mathscr{H}$ has the presentation

$$\langle S_1, \ldots, S_n | (S_iS_j)^{p_{ij}} = 1\rangle,$$

so $\mathscr{H}$ is isomorphic with $\mathscr{G}$.

Exercises

6.1 Suppose that $\mathscr{G}$ has a presentation $\langle \mathscr{S} | \mathscr{R} \rangle$ and that ψ is a function from $\mathscr{S}$ into a group $\mathscr{H}$. Suppose further that each relation in $\mathscr{R}$ becomes $1 \in \mathscr{H}$ if each $T \in \mathscr{S}$ that appears in the relation is replaced by $\psi(T)$ [loosely speaking we say that the elements of $\psi(\mathscr{S}) \subseteq \mathscr{H}$ also satisfy the relations in $\mathscr{R}$]. Show that ψ can be extended to a homomorphism from $\mathscr{G}$ into $\mathscr{H}$.

6.2 Suppose that $\mathscr{G}_1$ and $\mathscr{G}_2$ are finite groups with presentations $\langle \mathscr{S}_1 | \mathscr{R}_1 \rangle$ and $\langle \mathscr{S}_2 | \mathscr{R}_2 \rangle$, where $\mathscr{S}_1$ and $\mathscr{S}_2$ are disjoint sets. Show that $\mathscr{G}_1 \times \mathscr{G}_2$ has the presentation $\langle \mathscr{S}_1 \cup \mathscr{S}_2 | \mathscr{R}_1 \cup \mathscr{R}_2 \cup \mathscr{R} \rangle$, where $\mathscr{R}$ is the set of all relations $S^{-1}T^{-1}ST = 1$, with $S \in \mathscr{S}_1$, $T \in \mathscr{S}_2$.

6.3 Show that the modified procedure discussed on page 92 produces a word W' with one fewer partial words of length u.

6.4 Verify that the following statement provides a geometrical interpretation of Proposition 6.1.3: Of all the fundamental regions sharing a common edge, there is a unique one with the highest Roman numeral.

6.5 Show that the relation

$$S_1S_2S_1S_3S_2S_3S_2S_1S_3S_2S_3S_1S_2S_1S_3S_1S_2S_3S_2S_3S_1S_3S_1S_2S_1S_2 = 1$$

is a consequence of the relations

$$S_i^2 = (S_1S_2)^5 = (S_1S_3)^2 = (S_2S_3)^3 = 1$$

in the group $\mathscr{I}^* = \mathscr{I}_3$ of the icosahedron. It may be helpful to sketch paths on the surface of a cardboard model.

6.6 Show that the finite dimensionality of V is not essential in Schur's Lemma (Proposition 6.1.7).

6.7 Set $\alpha = -\cos \pi/m$. Show that the matrix

$$\begin{bmatrix} 4\alpha^2 - 1 & -2\alpha \\ 2\alpha & -1 \end{bmatrix}$$

has eigenvalues $e^{\pm 2\pi i/m}$.

6.8 Let $\mathscr{H}$ denote the rotation subgroup of a Coxeter group $\mathscr{G}$.

(a) Show that $\mathscr{H}$ consists of all elements of $\mathscr{G}$ that can be represented as words in an even number of fundamental reflections.

(b) If $T \in \mathscr{G}$ and $\mathscr{G}$ is irreducible, show that T can be written as $S_{i_1}S_{i_2} \cdots S_{i_k}$, where adjacent factors S_{i_j} and $S_{i_{j+1}}$ correspond to nodes of the Coxeter graph that are joined by a branch for all j.

(c) If a mark p_{ij} on the Coxeter graph of $\mathscr{G}$ is odd, show that S_iS_j is a commutator [e.g., if $p_{ij} = 5$, then $S_iS_j = S_jS_iS_jS_iS_jS_iS_jS_i = S_j^{-1}(S_iS_jS_i)^{-1}S_j(S_iS_jS_i)$].

(d) If $\mathscr{G}$ is irreducible and if every mark p_{ij} is odd, show that $\mathscr{H} = \mathscr{G}'$, the commutator subgroup of $\mathscr{G}$, so $\mathscr{G}'$ has index 2 in $\mathscr{G}$.

(e) In the cases not covered by part (d), show that $\mathscr{G}'$ has index 4 in $\mathscr{G}$ (see [9], p. 126).

6.9 Suppose that $\mathscr{G}$ is a Coxeter group.

(a) Show that there can be at most one element T such that $T(\Delta^+) = \Delta^-$.

(b) If $T \in \mathscr{G}$ and $T(\Delta^+) \neq \Delta^-$, show that $-r_i \notin T(\Delta^+)$ for some $r_i \in \Pi$. Conclude that $n(S_iT) = n(T) + 1$ (see the proof of Proposition 6.1.1).

(c) Show that there is an element $T \in \mathscr{G}$ such that $T(\Delta^+) = \Delta^-$, and conclude that $\mathscr{G}$ has a unique element with maximal length.

The remaining exercises require some preliminary discussion.

If $\mathscr{G}$ is a Coxeter group with fundamental reflections $S_1, \ldots, S_n$, then a *Coxeter element* of $\mathscr{G}$ is any product

$$S_{\pi(1)}S_{\pi(2)} \cdots S_{\pi(n)},$$

where π is a permutation of $\{1, 2, \ldots, n\}$. If $\mathscr{G}$ is irreducible and its Coxeter graph has no branch points, we agree to label the roots in Π in accordance with the following labeling of nodes of the graph:

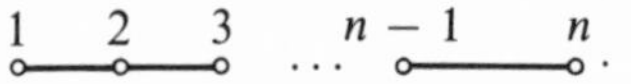

1 2 3 ... $n-1$ n.

If the graph has a branch point we label the roots in accordance with

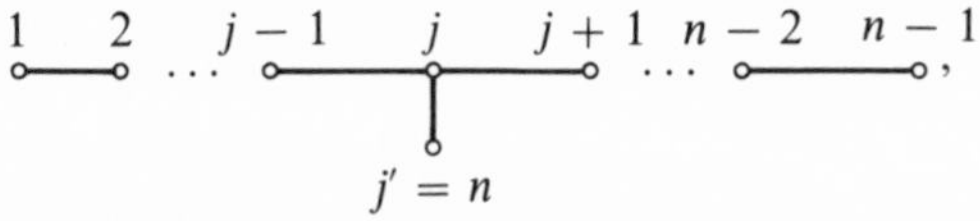

1 2 ... $j-1$ j $j+1$ ... $n-2$ $n-1$,

$j' = n$

where n is called j' to call attention to the fact that j' will be considered as *adjacent* only to the integer j among $\{1, 2, \ldots, n\}$, and in particular *not* adjacent to $n - 1$. In all other cases, adjacency has the usual meaning for integers in the set $\{1, 2, \ldots, n\}$.

6.10 Suppose that $\mathscr{G}$ is an irreducible Coxeter group.

(a) If the factors of a Coxeter element are permuted cyclically, show that the resulting Coxeter element is conjugate with the original one [e.g., $S_1S_2 \cdots S_n = S_1(S_2S_3 \cdots S_nS_1)S_1^{-1}$].

(b) If two adjacent factors with nonadjacent subscripts are interchanged, then a Coxeter element is unchanged.

(c) Use (a) and (b) to show that if adjacent factors with adjacent subscripts are interchanged in a Coxeter element, then the resulting Coxeter element is conjugate with the original [e.g., $S_1 S_2 S_{i_3} \cdots S_{i_n} = S_1(S_2 S_{i_3} \cdots S_{i_n} S_1)S_1^{-1} = S_1(S_2 S_{i_3} \cdots S_1 S_{i_n})S_1^{-1} = \cdots = S_1(S_2 S_1 S_{i_3} \cdots S_{i_n})S_1^{-1}$].

(d) Conclude that all Coxeter elements of $\mathscr{G}$ are conjugate with one another.

6.11 Show that all Coxeter elements of a Coxeter group $\mathscr{G}$ (possibly reducible) are conjugate with one another.

6.12 Find the *distinct* Coxeter elements in the groups of symmetries of the tetrahedron, cube, and icosahedron.

POSTLUDE

As indicated in the preface, the first systematic account of finite reflection groups was given by Coxeter [6] in 1934, following shortly after E. Cartan utilized the groups in his study of Lie groups and their associated Lie algebras. Coxeter classified not only the finite reflection groups, but also the infinite discrete groups generated by (affine) reflections. The infinite groups were likewise classified in terms of their Coxeter graphs. The relevant graphs are those that appear in Figure 5.4, with the exception of Z_4 and Y_5, and with the addition of $\circ\overset{\infty}{—}\circ$, which represents the infinite dihedral group generated by reflections in a pair of parallel mirrors.

There is an unfortunate disparity in the notations used to describe the finite and discrete Coxeter groups and their graphs. We have by and large followed the notation used in [4] by Carter, who to a large extent followed Witt [27]. Another variation appears in Bourbaki [3].

The finite crystallographic reflection group associated with a Lie algebra is called the *Weyl group* of the Lie algebra. The classification of irreducible Weyl groups gives a classification of simple Lie algebras over the complex field (see [4] or [21]), which in turn provides a local classification of Lie groups (see [18]). Each $\mathscr{B}_n$, for $n \geq 3$, is actually the Weyl group of two nonisomorphic Lie algebras called B_n and C_n. The distinction occurs because the crystallographic condition allows a choice in the assignment of relative lengths to the roots of the base, which results in two different invariant lattices.

We have followed Coxeter [6] in attributing the construction of fundamental regions in Chapter 3 to Fricke and Klein. Fundamental regions were perhaps first utilized in number theory some years earlier. In particular, they played a role in Dirichlet's study of the group of units in the integers of an algebraic number field. For an elementary account of the role played by fundamental regions in the study of automorphic functions see [23].

We have not included any discussion of the *invariants* of reflection groups, polynomials invariant under the action of the group elements. One example is the quadratic form $\Sigma_{i,j}(r_i, r_j)\lambda_i\lambda_j$. The notion of a Coxeter element (Exercise 6.10) is related. For a discussion of invariants see V.5 and V.6 of [3]. Also see Section 9.7 of [9], and [24].

It was indicated in Chapter 5 that Witt exhibited the root system of $\mathscr{I}_4$ as a group of unit quaternions. Coxeter showed ([8], p. 32) that the root system of $\mathscr{E}_8$ can be given as the set of units (which is not a group, but a *loop*) in a system of integral Cayley numbers.

The quadratic forms associated with A_1, A_2, A_3, D_4, D_5, E_6, E_7, and E_8 correspond to densest packings of spheres in dimensions 1 through 8. See [8], pp. 235–239, for a discussion and further references.

Finite reflection groups have come to play an important role in the study of finite simple groups. Among the earliest simple groups studied were the "classical groups," analogues over finite fields of simple Lie groups. Associated with each classical group is a reflection group, which may be viewed as the Weyl group of the corresponding Lie algebra, or may be constructed using certain subgroups of the group itself. A unified treatment of the classical groups, over arbitrary fields, was given by C. Chevalley in 1955, making possible the discovery of several new simple groups. Certain properties of the Chevalley groups were subsequently abstracted by J. Tits, who axiomatized the notion of a group $\mathscr{G}$ with a BN-pair, or a *Tits system* ([3], p. 22). One of his axioms is that $\mathscr{G}$ has a Weyl group, defined in terms of subgroups of $\mathscr{G}$ (as for the classical groups), which is a reflection group. Further details and references may be found in the excellent survey article by Carter [4].

Bourbaki defines Coxeter groups not as reflection groups but in terms of generators and relations. The point of Chapter 6, of course, is that the two definitions are equivalent. The proof of Theorem 6.1.4 might be viewed as an algebraic version of Coxeter's original geometric proof ([6], p. 599). Other proofs have been given by Witt [27], Cartier [5], and Iwahori [20]. The proof of the converse, Theorem 6.1.10, is patterned after the proof in Section 9.3 of [9].

Further historical information concerning reflection groups and related topics may be found in [3], pp. 234–240, and at the ends of the chapters in [7].

REFERENCES

[1] G. Birkhoff and S. MacLane, *A Survey of Modern Algebra*, 3rd ed., Macmillan, New York, 1965.

[2] H. F. Blichfeldt, *Finite Collineation Groups*, University of Chicago Press, Chicago, 1917.

[3] N. Bourbaki, *Groupes et algèbres de Lie*, chaps. 4, 5, and 6. Fascicule XXXIV, *Éléments de mathématique*, Hermann, Paris, 1968.

[4] R. Carter, *Simple Groups and Simple Lie Algebras*, *J. London Math. Soc.* **40** (1965), 193–240.

[5] P. Cartier, *Groupes finis engendrés par des symétries*, Séminaire C. Chevalley, Expose 14, Paris, 1958.

[6] H. S. M. Coxeter, *Discrete Groups Generated by Reflections*, *Annal. Math.* **35** (1934), 588–621.

[7] H. S. M. Coxeter, *Regular Polytopes*, 2nd ed., Macmillan, New York, 1963.

[8] H. S. M. Coxeter, *Twelve Geometric Essays*, Southern Illinois University Press, Carbondale, Ill., 1968.

[9] H. S. M. Coxeter and W. O. J. Moser, *Generators and Relations for Discrete Groups*, 2nd ed., Springer-Verlag, New York, 1965.

[10] H. M. Cundy and A. P. Rollett, *Mathematical Models*, 2nd ed., Oxford University Press, New York, 1961.

[11] C. W. Curtis, *Linear Algebra*, 2nd ed., Allyn and Bacon, Boston, 1968.

[12] P. Du Val, *Homographies, Quaternions, and Rotations*, Oxford University Press, New York, 1964.

[13] Euclid, *Elements* (3 volumes), translated with introduction and commentary by Sir Thomas L. Heath, 2nd ed., Dover, New York, 1956.

[14] J. Franklin, *Matrix Theory*, Prentice-Hall, Englewood Cliffs, N.J., 1968.

[15] R. Fricke and F. Klein, *Vorlesungen über die Theorie der automorphen Functionen*, vol. 1, Leipzig, 1897.

[16] M. Hall, *The Theory of Groups*, Macmillan, New York, 1959.

[17] P. Halmos, *Finite Dimensional Vector Spaces*, 2nd ed., Van Nostrand Reinhold, New York, 1958.

[18] S. Helgason, *Differential Geometry and Symmetric Spaces*, Academic Press, New York, 1962.

[19] I. N. Herstein, *Topics in Algebra*, Blaisdell, Waltham, Mass., 1964.

[20] N. Iwahori, *On the Structure of a Hecke Ring of a Chevalley Group over a Finite Field*, *J. Fac. Sci. Univ. Tokyo* **10** (1964), 215–236.

[21] N. Jacobson, *Lie Algebras*, Wiley (Interscience), New York, 1962.

[22] F. Klein, *Lectures on the Icosahedron*, 2nd ed. (translated by G. G. Morrice), Dover, New York, 1956.

[23] J. Lehner, *A Short Course in Automorphic Functions*, Holt, Rinehart and Winston, New York, 1966.

[24] L. Solomon, *The Orders of the Finite Chevalley Groups*, *J. Algebra* **3** (1966), 376–393.

[25] B. L. van der Waerden, *Die Klassifikation der einfachen Lieschen Gruppen*, *Math. Z.* **37** (1933), 446–462.

[26] H. Weyl, *Symmetry*, Princeton University Press, Princeton, N.J., 1952.

[27] E. Witt, *Spiegelungsgruppen und Aufzählung halbeinfachen Liescher Ringe*, *Abhandl. Math. Sem. Univ. Hamburg* **14** (1941), 289–337.

[28] P. Yale, *Geometry and Symmetry*, Holden–Day, San Francisco, 1968.

INDEX